Better Sound from your Phonograph
How come? How-to!
2nd edition

Robin Miller

author of American Radio Then & Now:
stories of Local Radio from The Golden Age

Better Sound from your Phonograph
How come? How-to!
2nd edition

an educational scientific & historic reference

Est. read time 16+hr (52,500 words plus hundreds of images)

by Robin Miller

editor R. A. Bruner

©2017, 2022 Robin Miller

ISBN: 979-8-218-06730-4

Updates & extras are downloadable at:
www.filmaker.com/papers/UPDATES_RMiller-Better_Sound_from_your_Phonograph.pdf

Unless noted, images and text are property of the author, All Rights Reserved. No part of this book may be reproduced in any form, except for brief attributed quotation in reviews, without consent in writing from the author or his agent. Tradenames incidental herein are property of their owners.

This book is for those who listen to music on phonograph records

("vinyl" on a turntable, record player, gramophone) who are curious about how they work, or why it is when they don't, even if they also listen to digital.

For those who would like expert resources that examines the science of a stylus rendering the very best quality sound that is baked into grooves of well-made recordings, mastered with precision lathes and pressed hydraulically in vinyl. Who need that knowledge to extract that *better sound*, that is engraved in grooves awaiting rebirth. Those who would marvel *How Come* a technology that is obsolete can possibly sound good as it does. For those who would get great satisfaction knowing *How To* select a better stylus profile, align an arm cartridge for least distortion, connect to a preamp through proper loading for optimal frequency response (tone color). Furthermore for the handy hobbyist who wants to save costs, this book offers this in technology you can make, or make better yourself! With maker instructions for a high performing yet low-cost preamplifier. A low-distortion transcription-length tonearm from ordinary hardware. This is a technical book for enthusiasts, new or experienced, illustrated in micro-photography posters & charts explaining the distortion mechanisms to avoid to enjoy your music. Sound quality at the end of a chain of audio links, with tips for amplifier power and loudspeaker sensitivity in the downloadable Update, but starting in print with a stylus that you will learn to choose for best sound, align for lowest distortion & wear, replace when needed, and enjoy in the meantime.

After reading, hold at some distance and squint to see a stylus contacting the groove walls at ✱.

LaunchPad

Contents

LaunchPad 2

Contents in pictures – a peak at what awaits you 6

Preface 2nd edition Science of "vinyl" and its rich history.................................... 8

Introduction 1st edition real sound, or fake? 9

The sound of "vinyl" is the sound of its stylus... 11

Phono styli close-up – how they work; how they wear 13

    Beginning with the most advanced – line contacts 15

    New needles, flawed or perfect, before any wear...................................... 20

    European audio magazines once proclaimed it "the finest cartridge" 24

    Comparing line-contacts by side (tracing) radius 26

    A robust line-contact stylus for broadcasters, DJs, & archivists 31

    Between line-contacts and sphericals in quality – the elliptical stylus....... 33

    The narrowest elliptical, that evolved into the line-contact....................... 41

    The lowly spherical (sometimes called "conical") needle........................... 42

Before microgroove, 60+yr of SP 78s & electrical transcriptions (ETs) 43

    SP (wide\coarse-groove) restoration\archiving styli.................................. 45

Phonograph styli faithful to the recorded groove 50

Vectorscope images of the stylus' movements 51

Stylus unfaithful to the groove – mechanical distortion mechanisms 54

Replacement \ aftermarket \ generic styli – caveat emptor 56

How the stylus & partners re-construct sound 58

    Alignment – a pickup cartridge mechanically mated with a tonearm......... 59

    Causes of "skating," and adjusting anti-skating (bias)................................ 62

    Cantilever-tonearm mechanical resonance... 63

    The stylus encounters a century and a half of dirty filthy disks 64

The "spinning" part of the turntable 66

    "Why bother, I can't hear it!"... 67

Recommended cartridge specifications & glossary... 68

Starter list of sources online for pickups, styli, info 69

Pickup \ preamp \ speakers – a system.. 70

"Tuning" the cartridge with the preamp 72

    Capacitive loading of moving magnet and moving iron pickups 74

Balancing two channels for mono & stereo groove replay 77
We've only "scratched the surface" of disk history! 81
Evolution of grooved disks: 78 rpm to 33⅓; SP to LP 83
Records "equalized" for lower noise, no groove hopping 85
Measuring sound (science alert) 90
The decibel (dB) & human listening 92
Sound Perception and better sound from your digital 95
Habituating to coloration (distortion), level compression, and no low bass 102
Gallery – "A picture is worth a thousand words." 104
A maker phono preamplifier with essential controls 107
Modify an A-310 PCB in five (5) check-by-steps 108
Wiring, setting up, and using the modified preamp 110
Connect the PCB to external equipment 111
Three ways to power the preamp 112
Avoiding hum by design 112
How to balance a cartridge's channels with a modified A-310 phono stage 113
Measured performance of the modified preamplifier 114
Conclusion installing the modified RIAA phono stage 116
A transcription tonearm – how come; how-to 117
1. Background & purpose - a high-performing 12in (305mm) phonograph arm 119
2. Bill of materials and initial machining (check off boxes as you go) 121
3. Fabricate wand sub-assembly 123
4. Wiring (up to preamp connectors, which are user's choice) 126
5. Fabricate finger lift; armrest & clamp sub-assembly 127
6. Tonearm installation on base, alignment, & testing 128
7. Fine tuning – optimizing arm resonance, skating compensation 131
8. Concluding construction 133
9. Steampunk tonearms "resonate" far & wide 135
Better sound from your entire audio system 138
UPDATES & extras (hyperlinked content) – www.filmaker.com/papers.htm 139
Grammy-winning mastering engineer Clair Dwight Krepps 140
Acknowledgements, the author, and editor 141
Index 142
Also by Robin Miller 145

Contents in pictures – a peak at what awaits you

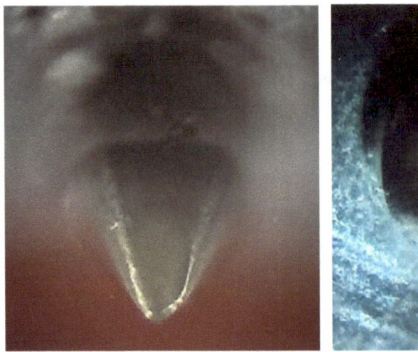 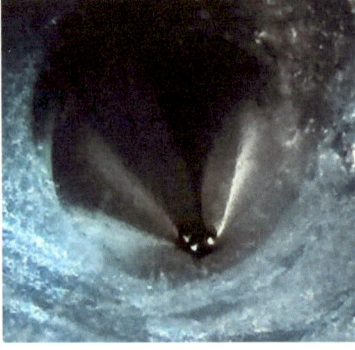

Left: How to tell a worn stylus damaging records - What you can evaluate with a $5 loupe.
Right: A fine line-contact diamond tip: The sound of the phonograph *is the sound of its stylus!*

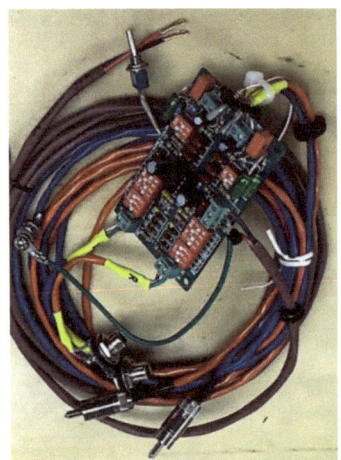 

L: Maker steps for a "phono stage" for ~$35. *R:* A UK reader's preamp nicely mounted in a metal enclosure, and reportedly working great after C-load selection and gain balancing.

A 1st edition reader's DIY low-distortion "steampunk" 12in (305mm) transcription-length tonearm of ordinary hardware from maker instructions. *Demonstrate its sound before you let anyone see it!*

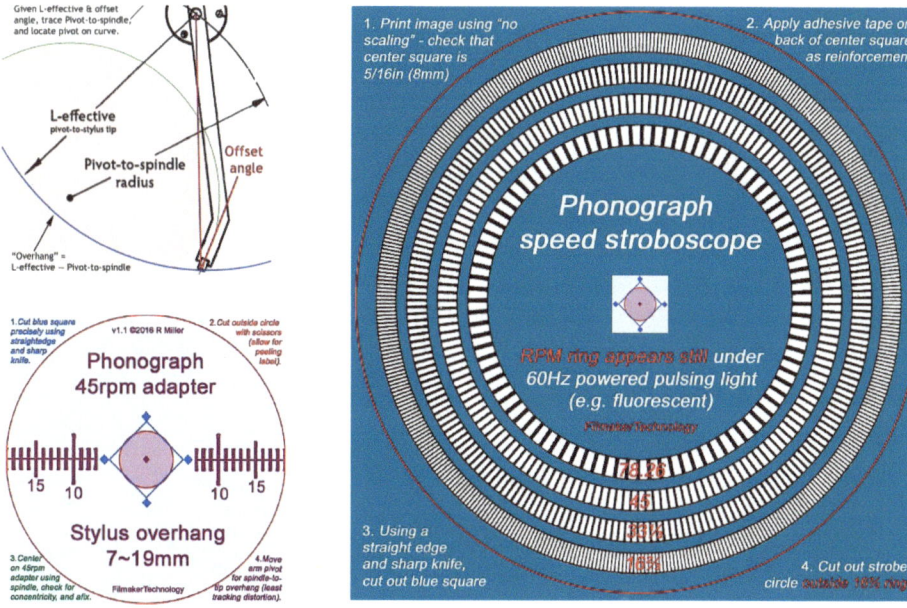

L top: Geometry behind tonearm alignment. ***L bottom***: lowest tracking distortion with this **stylus overhang** *checker* for attaching to a 45rpm adapter. ***R:*** Strobe disk checks turntable speed (musical pitch) – cutouts after downloading from http://www.filmaker.com/papers/UPDATES_RMiller-Better Sound of the Phonograph.pdf.

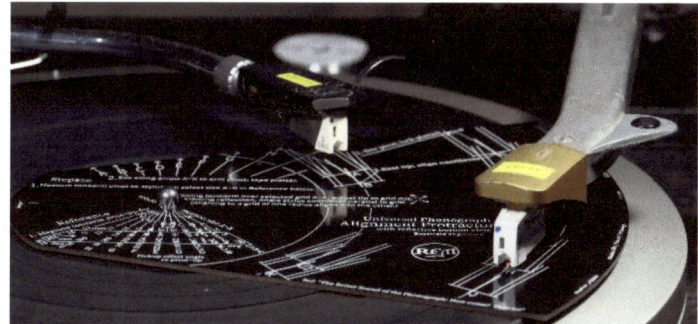

Above: Author's mirrored *alignment protractor* for cleanest enjoyment of music. ***Below:*** Digital over-processing squashes dynamics and **clips** wavetops, causing harsh distortion. Vinyl cannot be clipped, thus it preserves tone color. Recover the ultimate sound quality baked in a groove with user know-how.

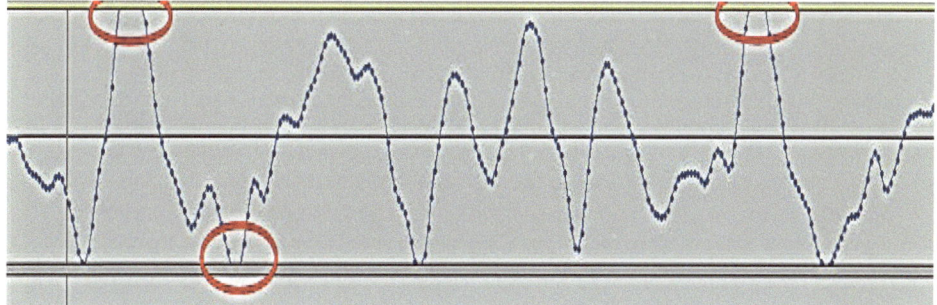

Now on with How come? How-to!...

Contents in pictures – a peak at what awaits you

Preface 2nd edition Science of "vinyl" and its rich history

Five years in, the book in your hand is new, all sections rewritten, content added, graphics sharpened, and an index added. Even the title is altered, as it got lost in translation: Technically "vinyl" is NOT better than digital....

I wrote so several times, but the original title caused misunderstanding. I only claim records sound better than digital over-processing and *clipping* that breed distortion. *Level compression* empties digital's least significant bits, reducing dynamic range to 12 *usable bits* (vinyl's equivalent). It gaslights shoppers to mistake immediately louder as better. Yet if you'd simply turn up play volume, recordings would be louder but also retain natural dynamics (loud v. soft), harmonic relationships (tone color,), and sound *better.* For many, over-compression artifacts are irritating. For others, artifacts are "brightness," possibly making up for deficiencies in their systems. Or the distortion becomes *habituated,* their *new normal,* though false – not *high fidelity!* Yet since the 1950s, analog vinyl still qualifies as high-fidelity.

Among many clarifications is the technical term "cutoff" in the chart on **p22**. It defines the *top* of any filter's cliff, not the bottom. Where the signal has only apparently fallen 3dB at a "cutoff frequency," f_3. But it's not the end of the world at the bottom, down several tens of dB. For "vinyl," cutoff is the *onset of loss* of high frequencies (HF). The chart on **p22** also speaks to the issue of vinyl v. digital. For example, a CD does not suffer in HF by end of disc – it has the same technical quality beginning to end, measurable in frequency response, distortion, and noise, the limitations of which engineers must heed mastering vinyl. Making test cuts, they check the sound with a "transparent" cartridge, finely aligned in its tonearm – the aim of this book for you*!* If trouble occurs while cutting the master, the side is scrapped and begun again. This diligence is one reason why vinyl now outsells CDs and downloads.

The new order follows the journey that begins with the engraved groove. For reader convenience, page referrals are in **green type**. Some readers may wish to save for later history stuff in **purple**, or higher-tech stuff **brown**. Some redundancy remains for readers' convenience, or for emphasis.

Again, the styli in micro-photo posters stand in for comparable products of other makers – it's not possible to include every tip by every manufacturer. They all share the science you will soon know in this book. In buying audio, seek honest specifications as cited throughout, and know how to compare.

Most digital devices are 'plug-'n-play,' but warn "No user-serviceable parts inside." *This book encourages turntable users to delve inside.* Because most anomalies can be minimized by readers, either hobbyists who like tinkering and science, or professionals who need optimal performance. By those who like music and other content on disk that will sound better after optimization. New micro-photo posters are key to choosing styli by performance and wear of your records and themselves. New context about other components in an audio system, and about who mastered the "loudest record ever recorded." For easy reading, the gnarliest math is in the footnotes. Let's turn the page…

Introduction 1st edition[1] real sound, or fake?

"Why do records sound better than digital?" As a sound conservator & audio engineer, I hear this question more than others: From inquiring turntable enthusiasts, and from the upsurge in curious newbies. And while *uncompressed digital reproduction is technically superior,* too much digital sound as distributed to consumers is subjectively inferior to analog vinyl's "warm" (connotes "natural") sound. Causing a resurgence in vinyl, new and used, in a market that exceeds CDs and digital downloads. Compared to that over-processed for retail, *vinyl sounds better – if played properly.*

This book is about playing vinyl *properly.* Probably better than you are now, or, if new to it, better than you might otherwise achieve. It presents findings from the author's endlessly asking questions himself, as a young hobbyist, then as a professional for 60+ years. Do you know records are a primary source of digital content? That *140+ years* of recorded history are archived on grooved media? That *degraded quality* is most likely due to improper playback? That typically the *needle rules how records sound?*

You might have jumped to the conclusion, in an mp3 world, that record playing is *passé.* I did. Yet I've returned to vinyl after decades working with analog magnetic tape and digital audio workstations (DAWs). And young people in my son's and grandsons' generations observe that, while image display has improved dramatically over the past 30 years, much consumer audio quality has over 30 years gotten worse. *Ear-fatiguing* in causes of distortion, fixed long-ago, but back today, and sounding as bad as long ago. Distortion reintroduced by marketing*!* Digital and analog audio need the pendulum to swing back. For the "record machine," the comeback is *now.*

And vinyl is coming back. *The Economist* [1st ed. 2017] reports vinyl sales tripled since 2010 [and nearly tripled again by 2022]. Demand for record pressing dramatically exceeds supply, with startups in Germany and Canada making new machines. LASER lathes to shoulder the load on the 25 mechanical mastering lathes in the world that still work. Whether online or in record and bookstores, more younger buyers than older buy an album on vinyl *after* streaming it, due in part to larger album art, and stories in text.

Many books have been written about the music and other historic content published on grooved media. This book has been written because *using phonograph equipment* is enjoying a renaissance as well. This book is not a shopping guide – online forums and sales websites cover the equipment, both vintage and the latest (hyped as "greatest"). From the pen of a degreed engineer (an applied scientist), this book is moderately technical. Analysis (how come?) precedes what to do (how-to).

For as wonderful a resource as it *can* be, the Internet, when it comes to audio (as for much else), is a haven for unsubstantiated beliefs rather than

---

[1] Edited.

proven facts. For pseudo-science – unknowing or intentional. The reverse-alchemy that would have you part with your hard-earned *gold* for poorer sounding *lead*. Not a book about conspicuous consumption, this is a quest for the best sound *and economy*. A book to serve the artforms of recorded music and the spoken word, but inclusive of those on a limited budget – one can always spend more. Whether for enthusiasts or just beginning, the author determined a need for a "textbook" apropos of a technological topic.

Worth "a thousand words" is each image, micro-photo poster, chart, and graphic illustration that clarifies how the phonograph works, why it sounds good (or bad), and what you can do to make it sound better. How to choose a replacement stylus (needle). With a soldering iron, make for $35 a high performing *"phono stage" (preamplifier)*. With a drill, how for the same $35 in ordinary hardware to make a low-distortion 12- inch tonearm. *Align* any tonearm for lowest distortion, and to minimize troublesome *resonance*.

What follows is about optimizing *transparency of sound* playing records. To be handy, it is a compact reference. Densely written to save the reader time, and the publisher trees, yet 80,000 words and hundreds of images including the online Update. For easier reading, math and techier stuff are in footnotes. For clarity, text observes the Oxford comma, two spaces following a period, and uniting numbers with their scientific units without a space. *Italics* introduces technical or foreign terms or lends emphasis*!* [2]

The author assumes you love music, enjoy recorded history, are techno-curious, and want better sound, not for background music, but for *focused listening* comparable to live performance. That you like vintage equipment and the *ritual* of operating it. That you find of interest a 21st Century update on the phonograph, invented ~1877. That you've attended to the basics, over-simplified in magazine articles titled "The 5 (7, 10?) easy ways to get *groovier* sound from your record player:" Placing the turntable on a solid level surface not near speakers; not touching record surfaces; not dropping the arm on disks; not playing 78s with a microgroove needle; etc. In this book, choosing a high performing, low wearing stylus, properly loading a cartridge, and aligning it with the tonearm are dealt with; plugging "interconnects" (audio cables) into the right holes is not.

Ultimately, this compendium is about *enjoying even more your investment in recorded music* and spoken entertainment on disk. By new research, expository posters, and working examples, the knowledge you gain to implement a phonograph system will extract the better full-range, distortion-free sound *baked-into* the grooves of well-made records. And add the satisfaction of maker solutions for an old turntable you already have. Or can easily find used. Your savings avoiding buying one wrong needle, or not attending to wear, could pay for this book many times over.

Vinyl's *achievable* sound – its proper replay – begins with the *stylus*.

---

[2] Metric units are in ( ) – for a fun look at Imperial units, see https://www.youtube.com/watch?v=iJymKowx8cY.

The sound of "vinyl" is the sound of its stylus

Because it is most critical in playing back (and wearing) the groove, we begin with a close-up exploration of gramophone needles (styli), under a laboratory microscope at up to 800 times magnification. Far exceeding the limits of a 60x inspection loupe (~$5 online with illuminator) or falsely advertised "1,000x USB microscope," this book's micro-photo posters reveal how *and how much the stylus affects sound*. How it also causes dangerous wear. At first glimpse of this previously unseen world, we don't know when to prohibit a needle's observable "defects" from touching our disks. More pages than you might expect will be devoted to understanding and minimizing stylus-produced distortion & wear. Simply because stylus' groove tracing is a far harder task than that of the electronics that follow.

Becoming aware of the workmanship on such a tiny scale, we learn what matters, and what doesn't. We see that needles come in many shapes and sizes, which make for as many different sounds. If they could be shaped like the flat-across stylus that cut the record, reproduction would approach perfection. But it can't be shaped like the *chisel* that carved the master, or it would re-cut the disk, erasing the details on it. With a cartridge capable of interchangeable styli, stylus selection can cover different record formats. But there are offerings to be avoided. Both used (worn), or brand new.

In contrast to a whole spectrum of things to know about the century-older electro-mechanical disk ["k" for records] player, the new digital devices are easy. Audio quality, whether in a sound card or on disc ["c" for CDs & DVDs] in a player, is tied to a key integrated circuit chip, the digital-to-analog converter – the DAC – and the steadiness of its *clock*. Consumer quality digital CD, DVD, SACD, DVD-A, and Blu-ray disc players simply work at their peak for years with little or no maintenance, if they do not suffer "infant mortality," or fail just after their 90-day warranty expires. If one functions past infancy or warranty, its main consumable part, its LASER, weakens slowly (over years), and at some point the player fails to track, and the music skips. In contrast, the turntable system works its best only if at the very outset it is properly assembled and aligned - enter this book. After which it serves with little maintenance, other than record cleaning and a drop of oil, for up to 1,000 hours until replacement of its diamond stylus.

In these first sections we investigate the science in graphs, charts, and micro-photo posters "suitable for framing" to understand the workings of these jewels. some of which come from chips after precious stone cutting, precision-ground to less than 1/10,000th inch (2.5 microns, millions of a meter, µm). Unlike a DAC, their workings are tangible, and aesthetic. But how to buy them amid the confusion of sellers' claims, after we know not only how differently they sound, but how differently they wear (destroy!) valuable records? Poorly made styli, or a worn stylus, tracking mis-adjustments, "skating," and disk un-cleanliness are the record groove menaces to know about. *How the best playing styli are also gentlest.*

Steinfeld says "the sound of the phonograph is the sound of its stylus," and this is largely true, but not the only factor. The sound as intended by the musicians, producers, & recording technicians is already baked into the disk, awaiting proper extraction by the stylus and its partners in the turntable system. The best thing the system can do in performing this process is to be *neutral* – not to wreck what's in the groove in any of several inherent but avoidable ways. Not altering the sound by adding noise and distortion caused by the stylus, electronics, or acoustics of the listening space. *How come* these anomalies exist, and *how-to* minimize these tone wreckers?

Prime examples of distortions caused by *stylus*-related issues are:
- *losing contact with groove* (misadjusted tracking force & skating);
- *adding distortion artifacts* from tip shape, misaligned tonearm tracking error, mis-tracing the groove, and tonearm resonance.

Several distortions sound similar, imprecisely called "sibilance" (because *vocal sibilants* are most annoyingly affected). Again, the distortions are likely NOT recorded in the groove, but are replay errors, are diagnosable, and are largely fixable by a savvy user with a useful book.

The stylus does not act alone. Vinyl's sound is affected by components that *complete* the turntable system: the stylus' own pick-up cartridge; the tonearm carrying the pickup; and the preamplifier ("phono stage"). Knowable mechanical & electrical interactions among these components, along with a consistently and quietly spinning platter, all affect the sound of the phonograph. If not bought as an integrated turntable package (low-to-medium in quality), these interactions require attention by an informed user\installer who expertly selects for purchase, then properly cobbles together, the turntable components.

Not the *sum of its parts,* any audio system is the *multiplication of its parts.* Any link in the chain proportionally affects the sound. After thoroughly investigating styli, first in the reproduction chain, we'll look at the partner components. Finally, the sound depends on listener *perception*, of years of habituation (conditioning) or *confirmation bias,* "fixable" by an open mind.

It is hoped the reader will find all this of increasing interest as the pages turn. But they are chock full of science and technical details. Too much to commit to memory, this book serves as a reference for any time one implements a new turntable system, as a guide for improving an existing one, and as a resource for when alignment or replacement arises.

The sound of the stylus is due largely to its tip shape, and the distortion it adds tracing a pristine groove. Like the stylus itself, *let us press on…*

Phono styli close-up – how they work; how they wear

When you want to help people, tell them the truth.
When you want to help yourself, tell them what they want to hear."
– Thomas Sowell

    To understand your choices of stylus by audio performance and by wearing of and themselves, we look at the most advanced tip shape before learning the limitations of more primitive stylus *profiles*. The ultimate profile is a *line contact,* explored after some context.

Good turntables with integrated tonearms are available used for less than $150 such as this direct-drive acquired on eBay. New units range $179 by U-turn Audio to $4,000 by Panasonic-Technics.

    The "Vinyl" (LP) is in a *renaissance* from its obsolescence and supposed extinction. Along with "shellac" 78rpm disks, the analog audio grooved medium of the phonograph (gramophone, turntable, record player) is a durable archive of nearly a century and a half of recorded music and spoken history. It has been surpassed in *technical quality,* first by magnetic tape, then by digital sampling. But due to to marketing-driven abuses of digital audio, records often surpass those in *sound fidelity* as typically delivered to the consumer. Over-processed mixing with level compression creates the illusion of sounding louder by squashing dynamics. By unnatural low frequency tone color from the relative raised levels of "normalizing." And by listening fatigue from intentional clipping distortion. For both popular music (largely electric instruments) and acoustic music (classical, jazz, etc.), the more natural dynamic sound of vinyl is confirmed in 21st century market data for used and new releases gaining revenue, and now exceeding CDs and lossy digital downloads.

    With help of resources such as this book, more users may make the effort to implement properly their turntable systems, align tonearms, electrically match cartridges with cables and preamplifiers, and *select a low-distortion stylus*. Micro-photographs explore how styli work, and how-to evaluate them critically for purchase, wear, and replacement. Users accustomed to digital media find the inconveniences of playing records can be offset by aesthetics and the satisfaction of optimizing replay quality. And the tactile ritual of gently inserting a gem stylus within a half mile-long spiral of sound engraved in a groove.

By the turn of the 21st century, I and other practitioners in audio – consumers, hobbyists and professionals – had pronounced "vinyl" dead. I did. I'd "recycled" (junked) turntables, arms, and cartridges from the 1970s. Fortunately, I kept a few. Then for me in 2012, a new client market emerged, requesting I restore 78rpm, *SoundScriber,* and radio ET disks they couldn't play. A professional audio engineer, my resume now included *sound conservator.*

By then I'd practiced mostly in analog tape and digital audio workstations (DAW), but disk reproduction had been *passé* for decades – so I thought. The next five years prior to publishing the 1st edition of this book was devoted to research catching up, realizing that I didn't know much in the first place, and I began to restore not only *disk* recordings ("discs" are optical media), but out of necessity, also the vintage machinery to play them. I began evaluating turntables and tonearms for repair, locating parts long out of production, or creating them anew. Seeking styli (needles), the main consumable of phonographs, but the best are nearly *unobtainium*; the worst the stuff of snake oil salesmen.

A record-playing system is an assembly of components: a cartridge & stylus (needle), tonearm, turntable ("spinner"), and preamplifier ("phono stage"). These are dealt with in the order the groove's undulations encounter them. Not the sum of these parts, the ultimate sound is a *multiplication* of its parts – any weak link will degrade *proportionately* the entire chain, ending at our ears. While these components are highly interactive, it is said about the potentially weakest link that "the sound of the phonograph is the sound of its stylus."

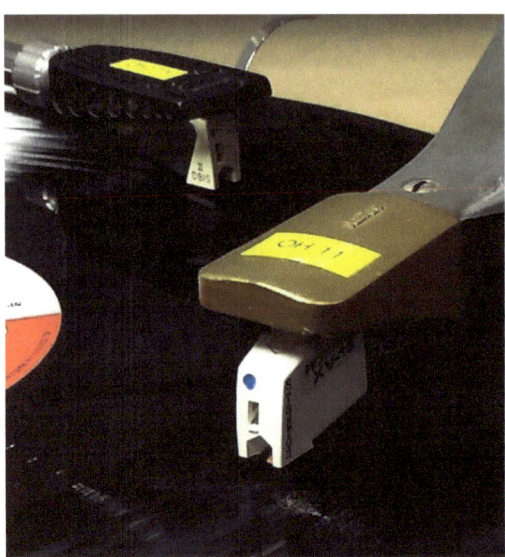

L: Two tonearms, two cartridges, two styli aligned for different disk formats. *R:* Interchangeable in a stylus grip.

Before examining at our closest the very tip, we look at the whole stylus assembly, in an interchangeable grip, illustrated above right, and in X-ray view **opposite**. Inside the grip's mounting tube, the *cantilever* is tied by a wire at the back and pivoting on a donut-shaped rubber-like fulcrum – an *elastomer* – that determine how stiffly the cantilever responds to a groove's undulations, the stylus' *compliance*. Low compliance for kid's record players; medium for disk jockeys; high for enthusiasts and archivists who gently handle delicate styli in low "mass" tonearms. At the back of the cantilever is a magnet or iron slug. When inserted its cartridge's body tube, the moving lines of magnetic force are picked up by coils that for this pickup are stationary inside the body. Other reproducers wrap the coils around the moving stylus tube and place the magnet in a fixed place within the cartridge body.

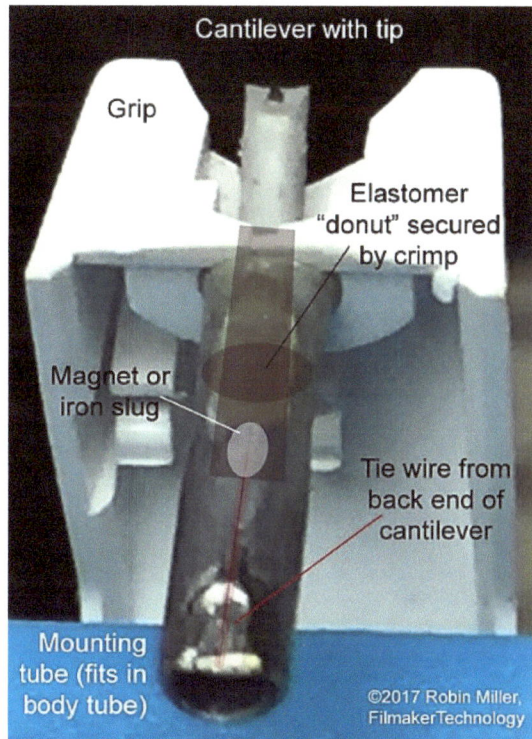

X-ray of a stylus assembly, the tip & cantilever are suspended in a donut-shaped *elastomer* with a *tie wire*.

Beginning with the most advanced – line contacts

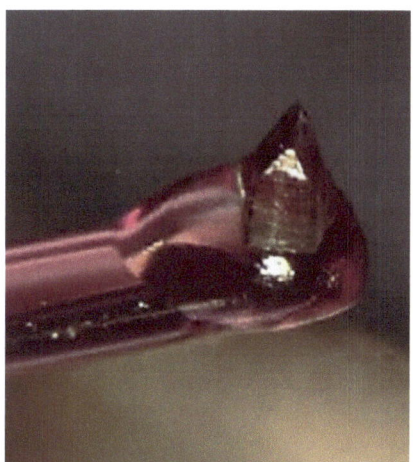 

L: Paratrace line contact tip in a sapphire cantilever. – Courtesy Expert Stylus UK. **R:** Electron microscopy tracing a vinyl stereo V-groove having separate L & R channel signals on each wall. – Courtesy Stanton.

Focused on one of the many phono cartridges (pick-ups) & stylus makers, herein by example, we'll deconstruct grooved media reproduction, aided by micro-photography. The challenge is that microgrooves are ~1/1,000 of an inch (1mil, or 25 microns, 25μm) across. Their contact regions are visible only under a lab-grade microscope. Needing miniaturized

Rembrandt modeling lighting.[3] High-definition photography. Multi-image stack focusing. Computer photo processing. In order to be interpreted in expository posters. Why? For evaluation. And for a semi-technical publication to help others, even new hobbyists, who may find them entertaining as well as edifying about a subject that is not dead after all.

For their quality and stylus interchangeability, our study focusses on *moving magnet* (MM) or *moving iron* (MI) "motors." Although the first pickups, later were reintroduced the more complex and expensive *moving coil* (MC) cartridges with fixed styli, lower compliance, and higher distortion. But we begin with a high-performing line contact stylus after an estimated 400 hours of LP playing time. The point (tip) is ground smoothest (gray in the poster below) where it has sunk in the groove to make contact. Higher, unpolished sides or a blob of glue affect nothing. Also illustrated in the frontispiece on **p3**, the tip is truncated so as not to bottom in the groove, where click-producing debris have fallen.

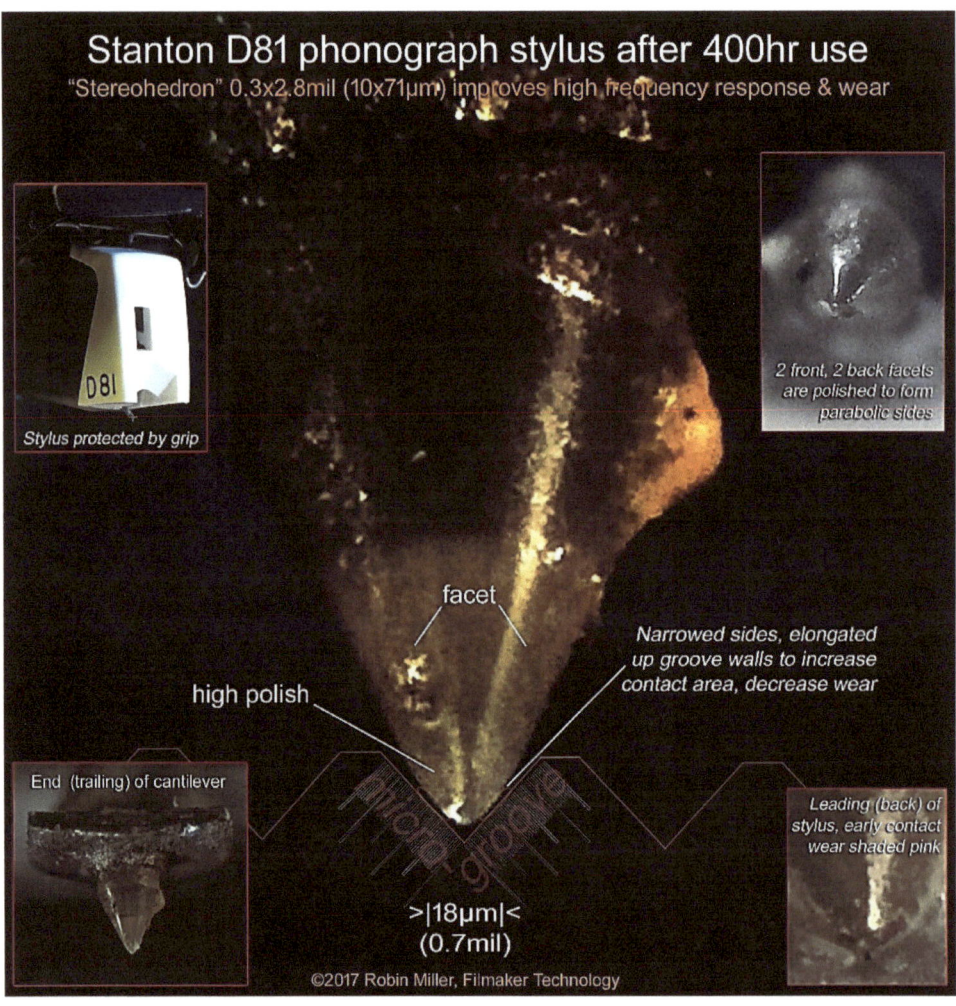

For a stereo groove, the contact regions, apart by 7/10,000in (0.7mil, 18μm, millionths of a meter, microns) are the most highly polished. In these regions the tip's critical side radii are 0.3mil (7.5μm) to trace its modulations in the direction of groove travel. Bearing on

---

[3] 30+years a cinematographer, I knew how to cast chiaroscuro light & shadow using luminaires 100~20,000w.

narrow "lines" of contact up\down both **V**-walls by 2.8mil (71µm) to spread the contact area that eventually wears both tip and groove permanently. As well as the life of records and the stylus, this shape affects the sound. Later we explore how problematically styli scan groove motion. And interfacing them with their pickups, tonearms, preamplifiers.

A stylus' inspection finds it in one of four categories: 1) a perfect specimen (I've yet to observe one); 2) an imperfect one but where adverse effects are negligible; 3) one begun wearing and increasing in distortion and losing high frequencies; and 4) damaging records.

Beyond Edison's primitive phonograph, who thought up modern teensy contraptions that extract high fidelity sound by dragging a hoe in a meandering furrow? Many developers were larger firms: ELAC and Ortofon in Europe; Western Electric, RCA, GE, and Shure in the US. One individual was Norman Pickering, whose styli are exemplified in this book. An engineer and violinist, he knew good sound. His partner\successor was Walter Stanton. All their efforts fill many patents & papers that with minor changes are still state-of-the-art.

Three tip shapes (profiles), detailed later, are the original *spherical* (some say "conical") that most consumers and DJs use, better performing ellipticals, and superior line-contacts. Their increasing costs tell us the extent of their grinding, drilling, polishing, orienting, and cementing, largely by hand. The meticulous work – under a microscope – was by artisans with steady nerves and inured to eyestrain. Then scrutinized by quality control. As said, we start with a line contact to reveal right away the ultimate stylus. Then clearer will be the limitations line-contacts addressed in ellipticals, and in turn the limitations of the original spherical shapes. Though on a micron scale, their significance is by no means small. Nor is their development and more than 10-fold improvement over a century in any way boring.

A disk mastering engineer would only use a spherical stylus to check how the disk would sound using a low-end needle. But for most work, many prized the D81 and line contacts of other makers. [Cf. *NB p34*] These high-end yet reasonably priced 1975 "Stereohedron" tips were developed by Pickering and successor Stanton Magnetics of Plainview NY. They improved upon the process of the first line contact by Shibata (1972 while at JVC), and preceded *SAS, MicroLine, FineLine, etc.* In this book, we advocate line-contact styli such as these because of their superior audio performance and low wear. Now scarce, many of us must settle for ellipticals. So next will come ellipticals, and lastly still useful sphericals.

Our next micro-photo posters show closer still the tip of two Stanton D81 styli, both top ranked line-contact "Stereohedron," both played ~400hr. This line contact – a translucent "nude" ground from a speck of gem dust collected from jewelers – is fitted through the flattened end of a tubular aircraft aluminum *cantilever*, securely bonded after aiming for the groove's approach from the back. But much higher magnification (up to 800x) is needed to observe formation and wear of the sides of its parabolic shield-shape that ride at 45-degree angles to groove walls as narrow patches, termed lines, that *trace* distortion-"free" sound.

At first, tips so enlarged seem rough devices, which a record collector would not want to touch his\her precious disk. These are not marketing "beauty shots." They present forensic evidence of both attributes and flaws. Don't be distracted by imperfections that make no difference to records or their sound – only the very tip's sides touch the groove. No need for manufacturers to polish or dab away a glue droplet far above a tip's contact depth.

Our concern, beyond manufacturing precision and audio performance, is ongoing *wear,* as shown by contrasting the next posters of two equally used D81s. They have approximately the same hours played, about half their life expectancy of 700~1,000h, the statistical mean before groove damage. After half expected life also begins a slow loss of high frequency (HF) sensitivity, and increasing distortion, especially evident by the inner groove. Also shown on **p16**, the D81 specimen next has life remaining; the tip on **p19** is near done!

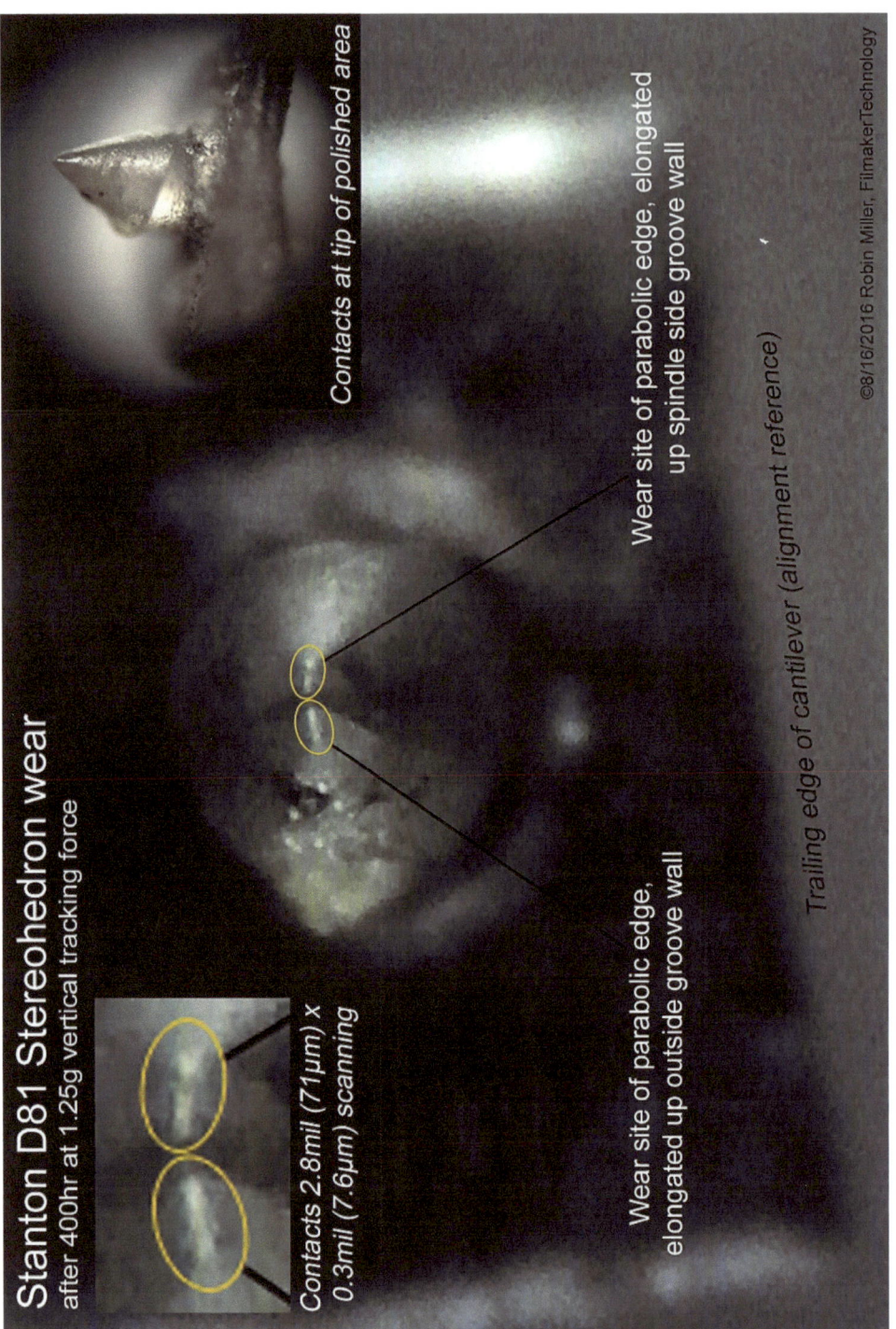

Stanton D81 Stereohedron wear after 400hr at 1.25g vertical tracking force

Contacts at tip of polished area

Wear site of parabolic edge, elongated up spindle side groove wall

Trailing edge of cantilever (alignment reference)

Wear site of parabolic edge, elongated up outside groove wall

Contacts 2.8mil (71μm) x 0.3mil (7.6μm) scanning

©8/16/2016 Robin Miller, FilmakerTechnology

1982 D81 "nude" diamond line contact stylus at 80x & 400x, well-aimed 2° off dead-ahead. Only the smoothest polished tip contacts the groove; the glue droplet up the cone is well out of the way. Groove-wall contact regions are white mirror reflections at either side that show minimal flattening wear after ~400hr; in the poster on **p16**, the straight forming at bottom left. As the tip approaches 1,000hr, these regions will widen, dulling HF response, and develop sharp outlines that do most groove damage. A gem in more ways than one, at half ideal life expectancy.

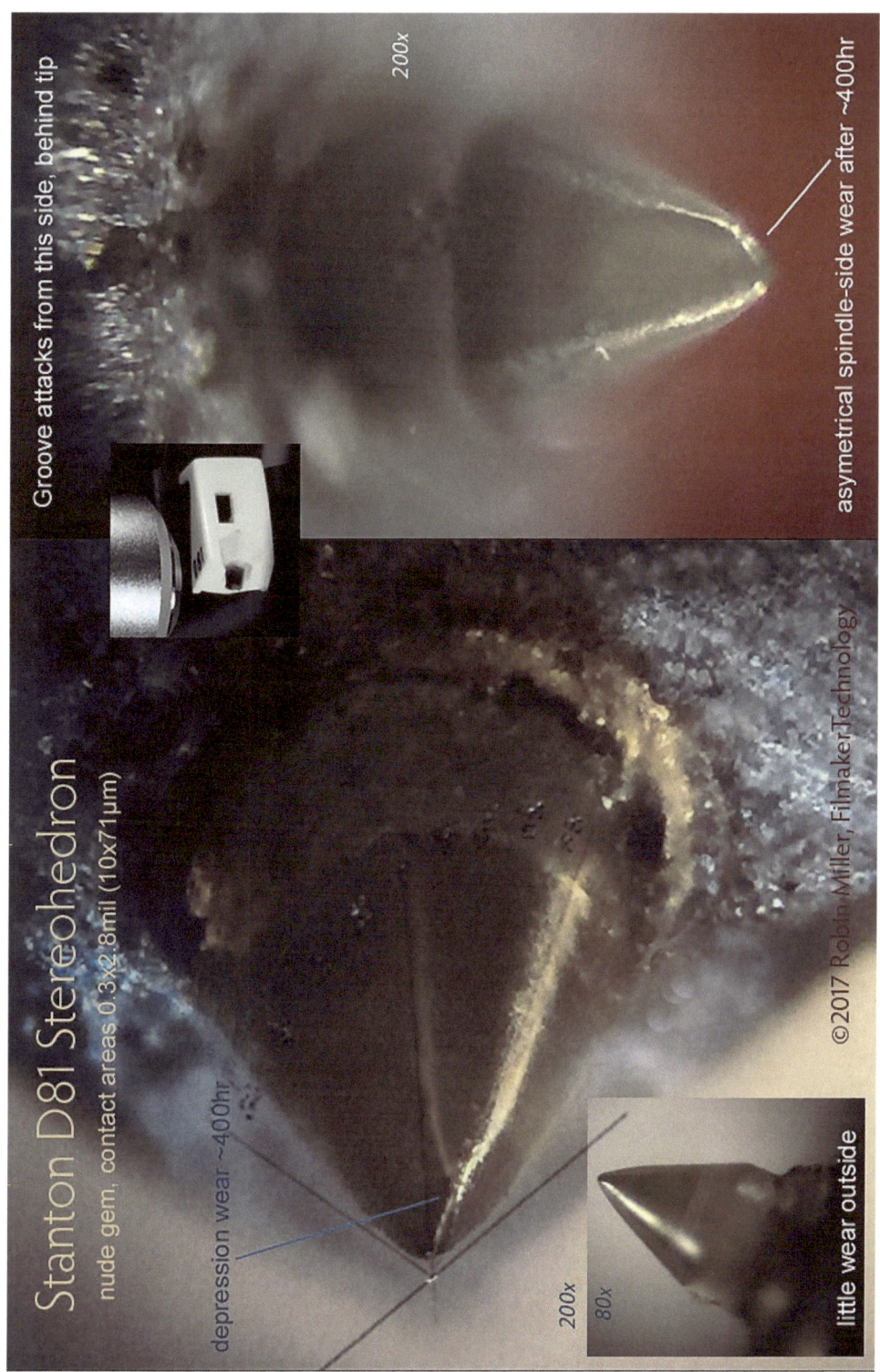

The 2nd D81 after hours similar to the prior D81. 90° protractor shows its points of line contact. A carved edge of the tip is toward the spindle, whittled by too little anti-skating. This stylus has begun to fail for fear of disk damage.

Phono styli close-up – how they work; how they wear

New needles, flawed or perfect, before any wear

The two D81 styli in the two preceding micro-photo posters show, after about half their expected life, wear that is still OK, and wear that is not. They can inform about poor alignment, or about estimating the future of styli that are used, new, or new old stock (NOS), specimens acquired from suppliers, or from prior owners at auction. Even having suffered "no wear," as yet, we discover flaws in manufacturing, especially lower value tips with looser quality control. But let's examine styli supposedly new\unused, in "as designed condition" right from the manufacturer, even if stored for decades. Simplest spherical styli are the easiest for makers to get right; ellipticals require precision facet grinding and alignment on the cantilever; line-contacts up the ante still further. Not just in making them, but also in aligning in arms by end users or their installers. But most critical is *tip shape*.

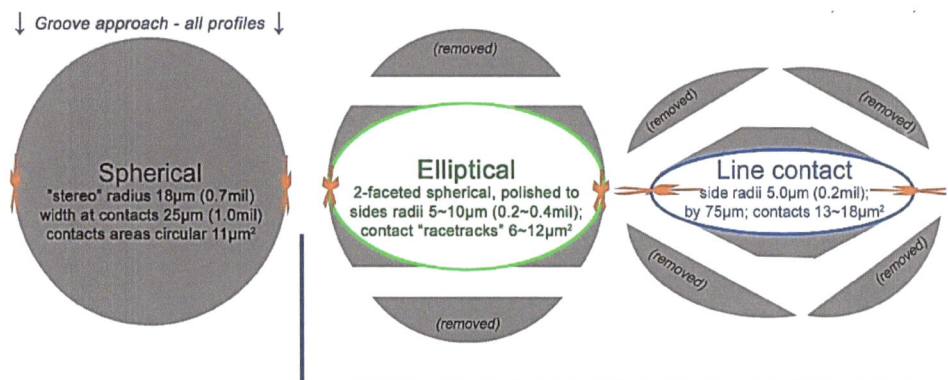

Microgroove stylus tips begin as hemispheric with radius 18μm (0.7mil), or between contacts ✶ 25μm (1.0mil). ©4/21/2017 Robin Miller, FilmakerTechnology

The drawings above show end-on the evolution in tip "profiles." From a single hemispherical radius, a second radius pair forming the sides of the elliptical, and a third radius pair bearing up & down the groove's walls forming the line-contact shape [Hughes' "improved Shibata" 1975 Patent US3871664 for Stanton Stereohedron]. Regardless of shape, it is only the sides that matter, at the **orange asterisks,** the only parts of the tip that touch the groove walls when the tip is immersed partway into the groove.[4]

Most economical to make and buy, the ***spherical*** tip is simply a rod ground to a cone, rounded at the tip with a 3D radius ranging historically from 0.5 to 8mil (12.7~200μm). Then it is rough-polished; the final polish may be in the first 10hr or so of the consumer playing his\her typical abrasively dirty records. Sphericals are bottom-line consumer or hurried DJ needles that produce the most distortion and least sensitivity at high frequency (HF), particularly nearing the end of a side. Predating ellipticals and line contacts, 78s and ETs for broadcast were played with wide-groove spherical styli, however the linear groove speed is significantly higher, extending HF response with lower distortion. Today generic styli replacements cost $25 on up from online providers, their provenance often unknown.

The lower linear speed microgroove LP eventually created need for a higher performing "hi-fi" stylus. The ***elliptical*** needle begins the same way as the spherical, a rod ground to a cone, the tip rounded to a hemisphere. Then flat facets are ground front & back at angles off vertical to form "race-track" shaped side contact areas. The more removed and subsequently polished, the narrower the sides, from *scanning\tracing* (side) radii of 0.4mil (10μm) to 0.2mil (5μm), perpendicular to each 45° groove wall at their contact patches.

---

[4] While cutting a master, groove depth & pitch are kept minimal, then increased in advance of higher audio level.

Ellipticals are truncated to avoid bottoming. And they must be aimed perpendicular to the axis of the cantilever and groove direction. Even the fattest tracing 0.4mil size reduces distortion and increases high frequency (HF) response, causing reasonable wear in middle-weight tonearms and tracking pressure. The narrowest 0.2mil size doubles the 0.4mil's high frequencies, and halves its distortion, but requires a lighter arm and pressure to avoid double wear. In between ellipticals of 0.3mil (7.5μm) side radii find many a home today.

Line-contact styli (Stereohedron, Microline, SAS, Shibata, FineLine, Paratrace, and other tradenames) are the most complicated to produce and expensive to buy. Posters that follow show four facets are ground, a third radius is established lying in the vertical plane of the contacts, and polishing is extensive. The goal of all line contacts is to shorten tracing as the groove passes for best HF response and lowest distortion, while lengthening the contact up\down the walls for greatest contact area to reduce wear of both groove and stylus. These find homes with professionals and hi-fi enthusiasts using manual turntables.

Drawings below dimension the tip profiles as viewed from front\back, bottom, and either side. All have tip width between contacts, 0.5~0.7mil (13~18μm) for stereo microgrooves. Tips contact the groove walls along their 45° angles. Ellipticals' side radii form short "racetracks" that momentarily softened the vinyl and indent into it due to heating at the contact area by tonearm pressure. Line-contacts add a radius pair originating far from the center of the tip. These sides are narrower, but taller racetracks. Elongation up\down the walls maintains contact areas to moderate pressure & wear. The line contacts must not be tilted, but be aligned parallel with the groove's HF by leveling the tonearm to set a vertical tracking angle (VTA, or stylus rake angle SRA), set between 15 and 21 degrees over time.

In nature, sounds statistically fall in level above ~2kHz, where the Recording Industry Assn. of America standard (RIAA, 1954) curve then has room to boost high frequencies (HF). As shown on the spreadsheet on **p22** beginning at frequencies termed "f_3 cutoff," HF reproduction is softer by 29% (–3dB). Then it falls 6dB per octave at ever higher HF. For high energy HF content, a 0.2mil (5μm) sided elliptical, Stereohedron II, and even narrower ridge-sided SAS or MicroLine tips extend HF response & transients, lower distortion, but raise wear. The chart on **p22** calculates (boxed in green) wall contact area, pressure, and relative wear cf. a D81 given the "reference" of 1.0, and "f_3 cutoff" by the inner groove.

Phono stylus dimensions	Point contact stylus				Line contact stylus	
	Conical stylus 0.7mil	Conical stylus (JICO) 0.6 mil	Conical stylus 0.5mil	Elliptical stylus (JICO) 0.3×0.7mil	Line contact stylus	S.A.S. stylus (JICO)
Front view	18μR	15μR	13μR	18μR	75μR	75μR
Cross-sectional View	18μR	15μR	13μR	6μR	6μR	2.5μR
Contact with records	L'3.8μ / 3.8μ	L'3.5μ / 3.5μ	L'3.3μ / 3.3μ	L'4.5μ / 2.5μ	L'9μ / 1.5μ	L'9μ / 1μ
Contact surface	30.5μm²	27.0μm²	23.4μm²	20.6μm²	46.7μm²	62.1μm²
L₁ / L₂	1	1	1	1.8	6	9

Dimensions of stylus tips: spherical ("conical"), elliptical (racetrack) of side radius 0.4, 0.3, or 0.2mil introduced in1964 by Shure., line-contacts of the 1970s, and even more exotic Super Analog Stylus (SAS ridge), courtesy JICO. Not shown, the monophonic microgroove 1.0mil "conical" until stereo (1958).

Phono styli close-up – how they work; how they wear

Groove & stylus wear v. high freq. "f_3 cutoff" Data Calc'd ©2017 Robin Miller 8/24/17d

Stylus mil	conWµm	conHµm	conAµm^2	VTFg	press.lb/in^2	speed	rel. wear**		f_3 kHz	***[Miller cf.*]
Sph 3.0x3.0	19.8	19.8	308	5	16,338	78	0.7	↑ if shellac	1.5	SP
2.0x2.0	13.2	13.2	137	4	29,408	33.3	0.5		1.0	radio ET 16in
1.0x1.0	6.6	6.6	34	3	88,224	33.3	1.6	HF erasure>	1.97	[*Goldmark]
.7x.7	3.8	3.8	11	2	177,425	33.3	3.1	HF erasure>	3.4	
.6x.6	3.5	3.5	10	1	104,572	33.3	1.9	HF erasure>	3.7	⎱ stereo range
.5x.5	3.3	3.3	9	1	117,631	33.3	2.1	HF erasure>	3.9	
Ellip .5x3.0	3.8	18.0	65	4	61,603	78	2.6	↑ if shellac	8.0	SP
.4x.7	3.0	4.5	12	3	260,854	33.3	4.6		4.3	
.3x.7	2.25	4.5	9	2	222,560	33.3	3.9		5.8	⎱ stereo range
.2x.7	1.5	4.5	6	1	160,477	33.3	2.8		8.7	
Line StHd	2.25	8.4	18	1	56,457	33.3	1.0	= wear ref	5.8	D81 .3x.7x2.8
DJ StHd	2.25	8.4	18	3	169,372	33.3	3.0	<2.8 at 2¾g	5.8	D6800SL "
.2 StHd ii	1.5	9.0	13	1	77,257	33.3	1.4	<1.8 at 1¼g	8.7	D81Sii .2x.7x3.0
.13 SAS	1.0	9.0	9	1	114,467	33.3	2.0		13	JICO .13x.7x3.0
.10 Quad	0.8	8.4	6	1	162,742	33.3	2.9		17	Pickg D4500Q

**varies w/VTF, speed, friction. Size data courtesy JICO, or interpolat'd/Miller.   ***Outer groove f_3 ~double.

Tabulated above are stylus profiles by wear of the tip and records it plays, HF "cutoff f_3," and implied distortion. The table informs choosing a new or replacement stylus. Tips have side scanning radii of 0.1~3mil (2.5~76µm) from LP to SP (standard play 78s & ETs). Fat ellipticals and sphericals or might do for some content, or for passing more quietly over dirt & scratches. But for clean, wide range content, sharper sided tips ≤0.3mil (7.6µm) are best.

The most referred to chart in this book, a deep-dive begins with its column headings:

Stylus mil categorizes spherical, elliptical, and "line-contact" cross-sections by radial dimensions: "tracing"\"scanning" along each wall x "bearing" wall-to-wall, in mils (1/1,000in, =25.4µm);

conWµm is contact width W along each groove wall, the groove wall indentation in microns (µm);[5]

conHµm is contact height H (inclined 45°), the indentation up-down each wall, in microns (µm);

conAµm^2 calculates each side's contact area A, indented in the groove wall, in square microns (µm^2);

VTFg is the tonearm's set vertical tracking force pressing the stylus into the groove, in grams (g);

press.lb/in^2 is the share on each wall at 45° of instant pressure from the VTF, in pounds per sq in;

speed is the record speed in rpm, either 33⅓ or 78 [Groove speed about halves across the disk.];

rel.wear estimates a factor for wearing of both stylus and record grooves (due to tip shape, VTF, friction, & speed) relative to the low-wearing D81 line-contact stylus as a reference, assigned the value 1.0;

f_3 kHz is the so-called "cutoff" frequency above which a given shaped stylus can no longer trace the inner groove at maximum modulation (volume). A cautionary "speed limit" at the dawn of Hi-Fi. [6]

Perhaps too strong a word, "cutoff" is standard engineering terminology for the *onset* of signal filtration – the start of descent of a hill, not the bottom of a cliff. "f_3 cutoff" is also applicable to the sensitivity v. frequency of speakers, mics, etc. For phonograph pickups, filtration is caused by the physical limit of stylus tracing – aka its "groove curvature overload." At f_3, that frequency's response is reduced 3dB, causing an audible but fleeting

---

[5] "conWµm" is less than the side radius, but the approx width of each contact's softened, indented "racetrack."

[6] The LP's 1,970Hz "cutoff" (–3dB point) using a 1mil spherical [Bachman (Goldmark), Columbia 1948].

reduction in brightness at highest level. It's not the end of the world – there are 60+dB to go! The inability of a stylus to trace a high frequency groove instantaneous mellows the peaks of brassy trumpets, crashing cymbals, triangles, fuzzy guitars, etc. It occurs most at the innermost cut of a pop album, or the climax of a symphony. Competent mastering engineers make test cutting(s) of the level peaks of a side to ensure no deleterious effects.

In the chart, data are entered in yellow boxes; the results are calculated boxed in green. *Relative wear* (compared to a D81, given "1.0") combines calculated contact area *conAμm²* by the side of a stylus at each wall, record speed, and the vertical tracking force "VTFg" in round values. For a fractional value one actually uses, such as 1¼g or 2¾g (in this book as optimal for SL & Sii styli), just interpolate its relative wear. The high ***press.lb/in²*** heats and indents each vinyl wall (solidifies seconds later). The chart notes when a stylus may erase a groove's HF peaks, and the double stylus-wearing friction effects of shellac at 78rpm.

Of greatest interest with LPs in **braced } green** is the hi-fi "stereo range." The results show tradeoffs of high frequency response (HF) v. wear. **Blue f_3** calculates the worst case inner groove onset of HF loss at maximum level (double it for the outer disk) that are much improved since the mono 1.0mil spherical stylus introduced with the LP, with its "1.97kHz cutoff" [Goldmark 1948]. Sharpening the side scanning\tracing radius from 1.0mil (25μm) to an elliptical 0.2mil (5μm) improves –3dB cutoff to ~9kHz, but at nearly double the wear. For higher frequency, higher level content, a mastering engineer backs off the lathe cutter drive, linearly increasing the maximum HF any tip shape can reproduce. E.g. a 6dB lower modulation doubles f_3, but lowers playback by 6dB, worsening signal-to-noise ratio 6dB.

The rightmost columns, **wear in red** v. **f_3 in blue**, are the chart's essence. As examples:

- A 0.3x0.7mil elliptical tracking at 2g is characterized by an onset of attenuation at peak levels at about 6kHz, but has nearly 4 times the rate of groove and stylus wear as the reference line contact with the same cutoff, the D81 tracking at about 1g;
- Also tracking at 1g, a hyper- elliptical 0.2x0.7 that begins to mellow peaks above 8.7kHz has 3 times the rate of wear, but does no harm for rock albums and 45's if they were rolled-off above that frequency, as they typically were in the 1970s~80s.
- The roundest 0.4x0.7 elliptical cuts off above 4.3kHz tracking at 2g with about 3 times the D81's rate of wear (interpolated from 4.6x, listed for 3g), but is fine for piano music that, with its ~7kHz maximum spectrum, will likely not ever be so loud.

Cutoff mattered less in the phonograph's first 70yr, with spherical needles on rock-hard shellacs tracking in ounces (1oz =28g)! Speedy, hard surface, high friction 78s evolved in HF from 5 to 7 to 10 kHz. They wore out a diamond in 200 hours; popular sapphires in 20! A steel needle change was advised for every side! But then a 78 side lasted only 3~4½min!

In 1948 with the introduction of the LP, releases were still mostly orchestral, big band jazz, opera, and "nature" sounds. These acoustic sounds naturally diminish in energy level above ~2kHz that permitted a boost of HF. From 1953, RIAA standard pre-emphasis above ~2kHz is reversed upon replay to provide flat response along with reduced noise.

At about half the linear speed of the outer groove, the inner groove crowds the tight turns of high level HF undulations, technically termed "curvature overload," manifesting as what "audiophiles" term "sibilance" (derived from the raspy distortion of vocal sibilants). At frequencies above a tip dimension's inherent f_3, its side tracing\scanning size begins to be too large to fit the curves. The groove's midpoint becomes the playground for high level HF *pinch effect* distortion. By a side's end, a too-wide spherical tip has cut corners, digging into pressure-softened vinyl. In a few plays, inner content brightness can be "erased" permanently. These limitations keep conscientious mastering engineers alert.

Phono styli close-up – how they work; how they wear

European audio magazines once proclaimed it "the finest cartridge"

Ultimately, the line contact shape came about to solve the high frequency challenge of quadraphonic surround sound on disk, called CD-4, that *enveloped* the listener in 360° 2D, preceding by a decade 5.1 surround sound on DVD. Quad on vinyl needed 50kHz response for the subcarrier of a second signal pair. The solution was a super-narrow line-contact tip.

Just in time for Quad's demise in 1975, Pickering's D4500Q "Quadrahedron" CD-4's most successful stylus. Four deep facets produced a 0.1mil (2.5μm) tracing tip on a strong aircraft aluminum cantilever, with a costly sintered rare earth rod of magnetically powerful Samarium-Cobalt (SmCo) that generated high output. Even tracking at ½g it inflicted wear that "erased" ultrasonic nooks & crannies in Quad's groove within a few plays!

Three posters that follow examine closely the historic Quadrahedron stylus. Its bearing radius is 28x its very narrow width. Wear is compared in the spreadsheet on **p22** to 0.2 or 0.3 line-contacts (bearing radius 9~15x tracing). Its f_3 is 17kHz is 5 times a spherical's cutoff. By most closely emulating a cutting chisel, its tracing forms of distortion are 80% reduced! Due to its high wear, its use playing audible-band stereo is limited to pristine archiving (vinyl "ripping"). Long out of production, it is quite scarce. But Quadrahedron's principle would be adapted as Stereohedrons with wider scanning sides and lower wear.

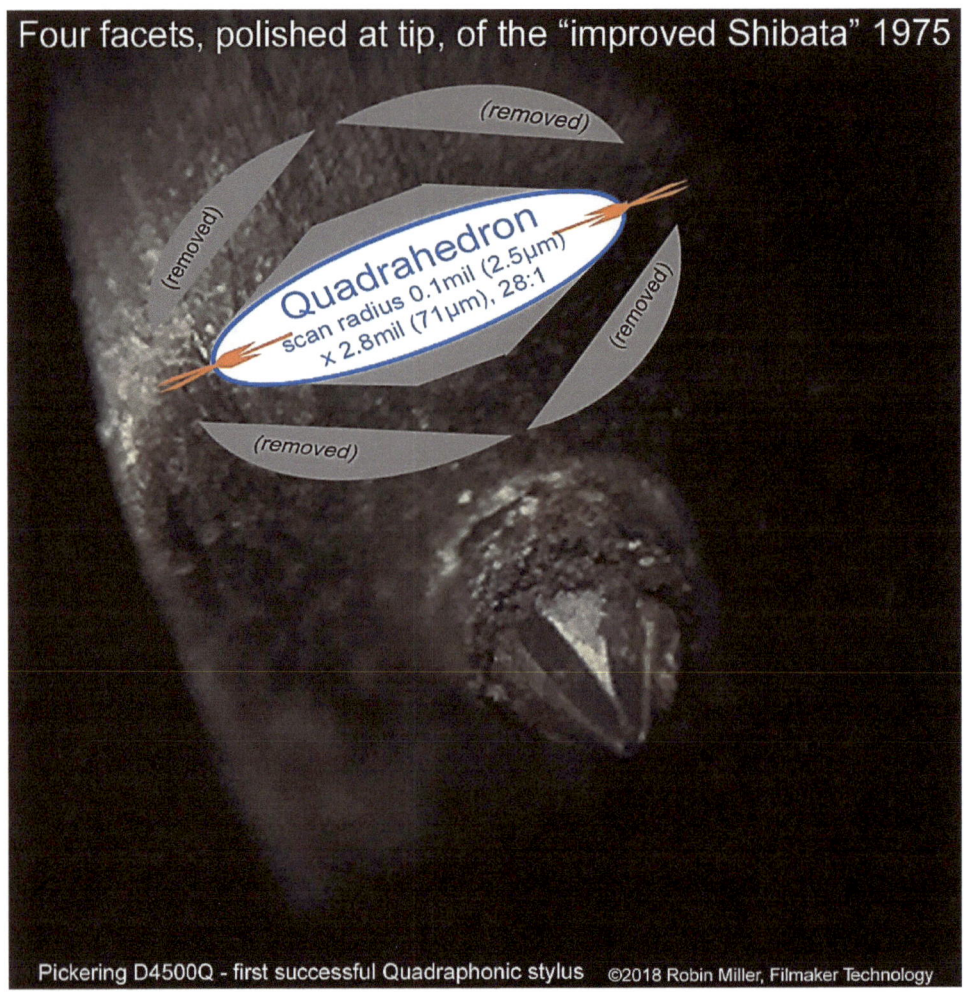

Pickering D4500Q - first successful Quadraphonic stylus ©2018 Robin Miller, Filmaker Technology

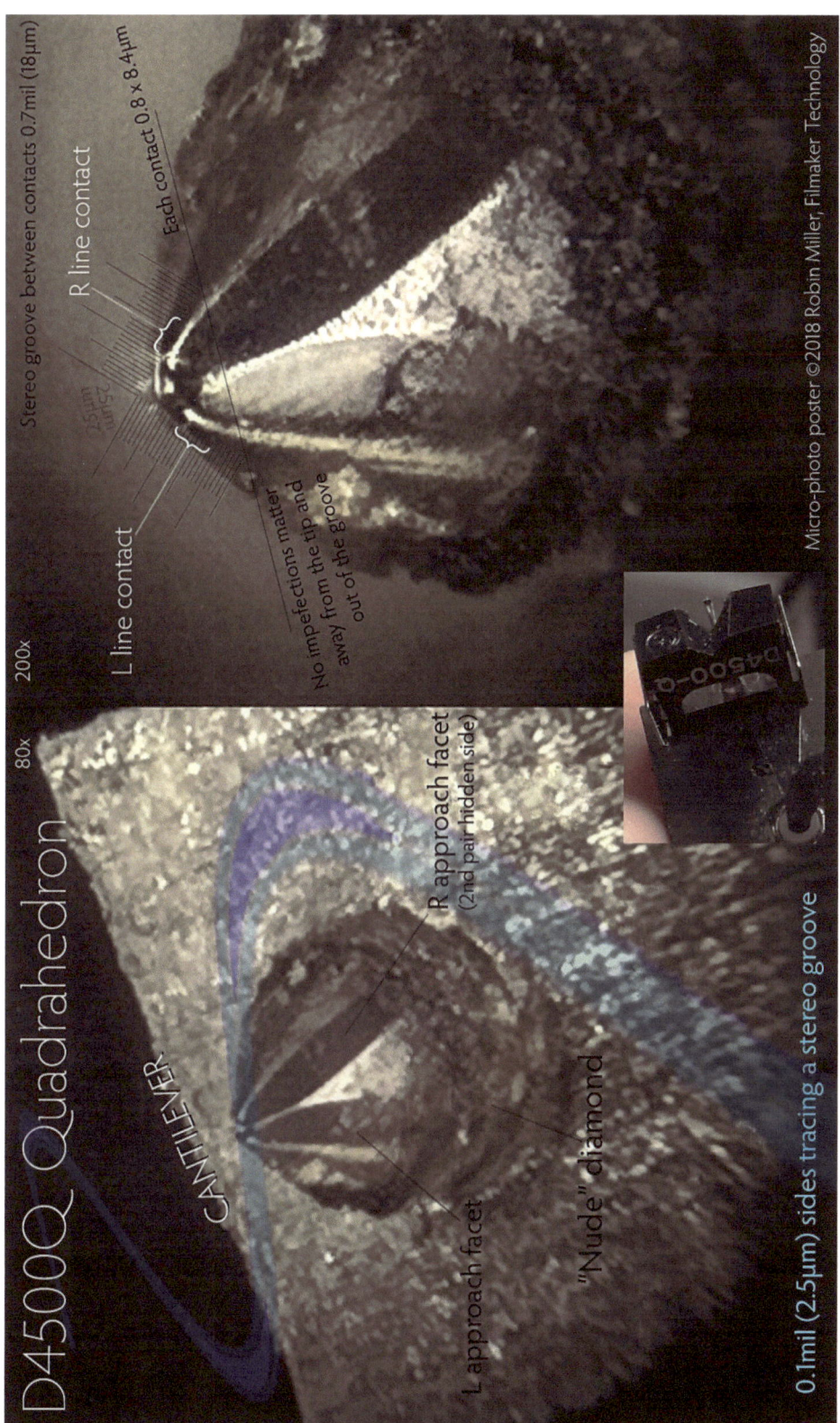

Phono styli close-up – how they work; how they wear

p25

"Line contact" in this book encompasses the evolution to develop a stylus capable of low distortion response to 50kHz for Quadraphonic surround. The market for Quad failed, however the implications for stereo were realized and are the state-or-the art still today.

The two posters above and the next illustrate this evolution. Shibata at JVC had in 1971 invented the line-contact stylus profile. Then in 1975, Stanton developed the "improved Shibata" that significantly reduced VTF and wear, simplified fabrication, and reduced costs. Pickering dubbed it *Quadrahedron,* a D4500Q specimen in the middle **opposite**. Its deeply ground facets, indicated by the long vertical leg of the "Y" shape between facets, result in very narrow side contact areas that extend f_3 to 17kHz, and a useful response to 50kHz at – 10dB below the audible range. The next year it was reintroduced for stereo, first as a 0.3mil (7.5um) *Stereohedron,* **opposite** at left, with f_3 of 5.8kHz that is a touchstone of this book for other stylus profiles. Later, **opposite** at right with f_3 of 8.7kHz, is the in-between 0.2mil (5um) Stereohedron II. The relative wear and f_3 frequencies are in the table on **p22**. Others' followed: FineLine, MicroRidge, MicroLine, VividLine, Paratrace, SAS, etc.

Narrowest shapes deliver flattest response and lowest "sibilance" distortion. A price can be more clicks & pops. Perhaps not needed as much for a soft piano album, line-contact styli are unsurpassed tracing raucus electric guitars, an acre of violins, or a bevy of soprano choristers. Along with a tiny summarium cobalt (SmCo) magnet, these achievements drew praise in Europe in 1976 for Pickering\Stanton models.[7] But selling through discounters caused backlash in the US hi-fi press, sent echoing among their audiophile devotees.

From its 1975 Patent's title "An Improved Shibata" realized first in Pickering D4500Q, the Quadrahedron, the successor 0.3mil Stanton D81 Stereohedron line-contact was among the finest of styli, though reasonably priced. Tested in each MM 881 pickup and sister MI 681 as a D6800eeeS tip, archivists & mastering engineers could rely on their individual "calibration" sheets. Many chose it for quality control, spot checking an LP while being lathed, so savvy users playing with the same pickup came closest to the sound intended.[8]

50+ years on and no longer made, NOS Stereohedrons have become scarce. For most of the many the author has tested, their SmCo magnets show no sign of weakening. However in some the elastomers have stiffened a bit, causing a modicum of reduced compliance from their original, quite high 30CU. Typcally less by 10~15% mostly affects resonance, easily compensated by raising tonearm mass. By far, most still measure well and sound excellent.

The author laments for his readers & clients that Stereohedrons are scarce. However their legacy exists new in tradenames above, or through a re-tipping service such as Expert Stylus UK, although the author has not tested them all. Most suppliers today compromise users' experiences by selling less costly ellipticals, with the consequences implicit on **p22**.

Comparing line-contacts by side (tracing) radius

Styli are nicknamed according to three radial dimensions. Their 1st radius is each side (tracing). Their 2nd radius is the standard width of stereo microgroove contacts, 0.7mil (18μm) wall-to-wall. By the 1st of their two radii, ellipticals are 0.4, 0.3, & 0.2mil (10, 7.5, 5μm). With less wear by 1st radius size, line contacts are 0.1, 0.2, & 0.3mil (2.5, 5, 7.5μm). Line contacts' 3rd radial dimension originates well outside the tip's sides, defining the patch of contact up & down each wall. This larger area distributes stylus pressure for lower wear.

---

[7] Pickering\Stanton models include XSV-3000~7500, 880\881 high impedance and 981 low impedance, and Walter Stanton's namesake WOS-100\CS-100, all having the reference D81 tip in the chart on **p22**.

[8] Note: "replacement" tips designated "D6800eeeS" are NOT genuine, nor true "line-contacts," as shown later.

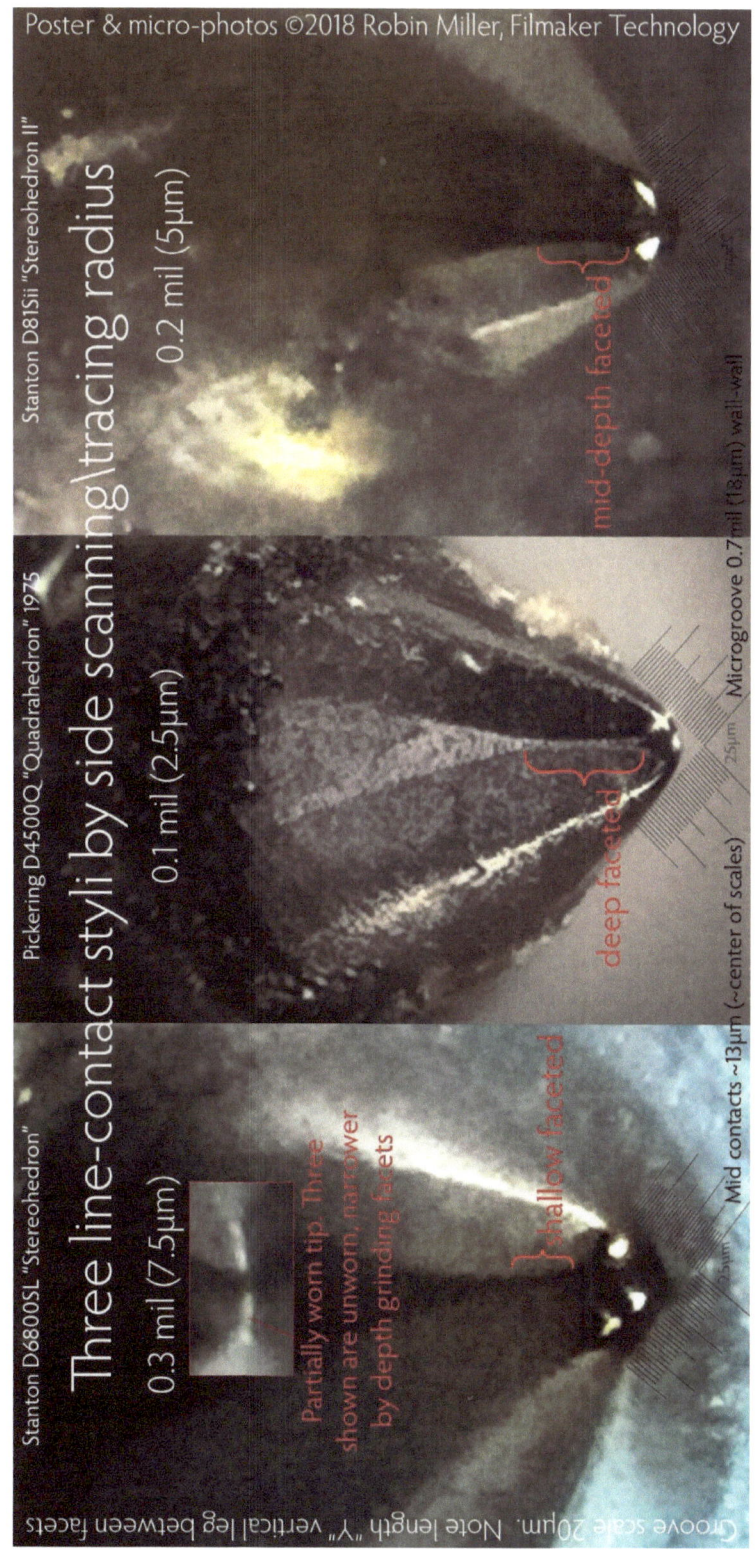

Phono styli close-up – how they work; how they wear

As said, sacrificing some wear for even better HF response, Walter Stanton, who had assumed Pickering management and renamed the company, introduced a ⅓ narrower 0.2mil (5µm) line contact. In the Wear & Cutoff table on **p22**, reducing side tracing raised a worst case f_3 cutoff of +12dB peaks in inner grooves of an LP to ~8.7kHz, implying ~17.4kHz in the doubly fast-moving outer groove. Also elongating the 3rd radius up the walls to 3.0mil (76µm) made up some lost contact area. At the same tracking force, wear is nearly the same as a popular 0.4x.7 elliptical. This "Mark II" in the poster below is the D81Sii.

A Stereohedron II, with its narrow scanning contact patch of 0.2x0.7x3.0mil (5x18x76µm), traces high level HF to f_3 of 8.7kHz in the region of the inner groove, where content often climaxes. Or 17.4kHz at 6dB below maximum level, or at outer groove at full modulation. Yet wear is lowest among hyper-ellipticals because its bearing area is taller up each groove wall, easing pressure.

The micro-photo poster above, the prior poster, and the excellent frequency response on **p30** tell the stories of D81Sii specimens certified by the author for archiving work. Using a CBS STR140 pink noise test disk, within its range of 30Hz to 15kHz, the response of the complete high performance turntable system with this stylus is within ±½dB, including errors of a custom RIAA preamp that independently is flat ±¼dB, and the trusted response of this test disk model (I compare three). The smooth, flat frequency response (FR) means the sound of the measured component will likely be accurate in tone color (timbre) – the holy grail of high fidelity. And while quantitative data do not always jibe with hobbyists' subjective opinions, this specimen is, prior to measuring, the equal of the best clean-sounding phonograph reproduction the author has heard. It sets the goalpost – for now!

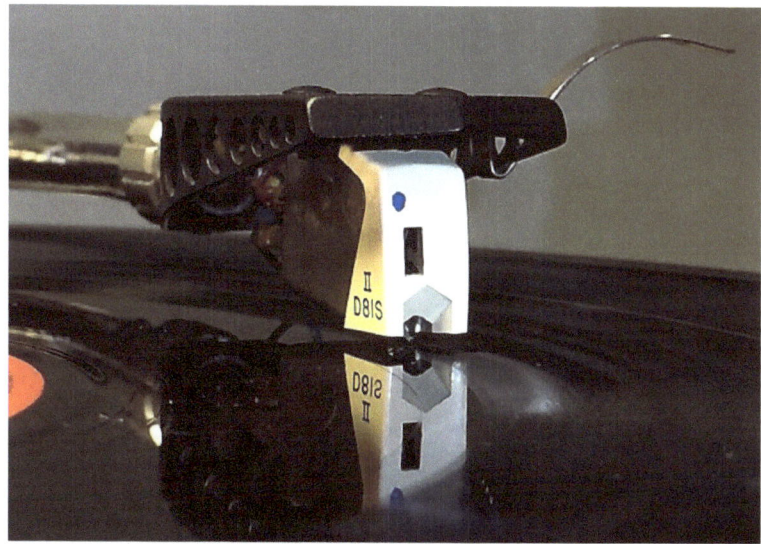

A Stereohedron II (dust brush removed) on an SME tonearm. Again to remind readers that, important as they are keys to the sound of the Phonograph, these styli from a pioneering maker are presented as examples that in time most makers emulated. It would be impossible to chronicle them all.

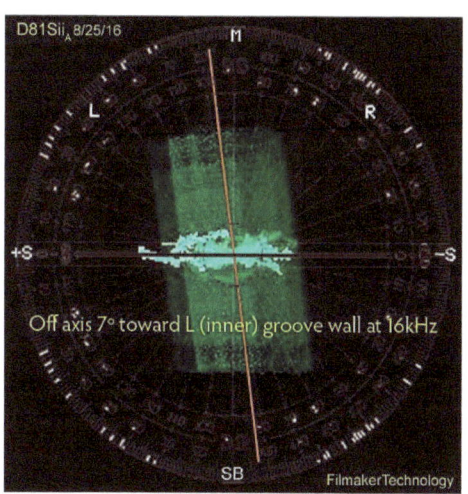 

L: Blue's tip is 7° off course (channels are barely visible off-phase on **p30**). **R:** A 2nd D81Sii is off-center (ignorable). Both manufacturing errors are borderline in QC; far more common in lower-end styli.

"Line-contact" styli by any maker have a narrow *bearing* patch approaching a straight line *up\down the groove wall* ranging up to 15 times the dimension of its scanning (tracing) sides *along the groove.* This 0.2 x 3.0mil (5 x 75μm) tip heats, softens, and indents vinyl less than the same 0.2-sided elliptical, causing less wear to both groove and stylus by spreading the contact pressure over a vertically elongated footprint, as tabulated on **p22**.

Phono styli close-up – how they work; how they wear

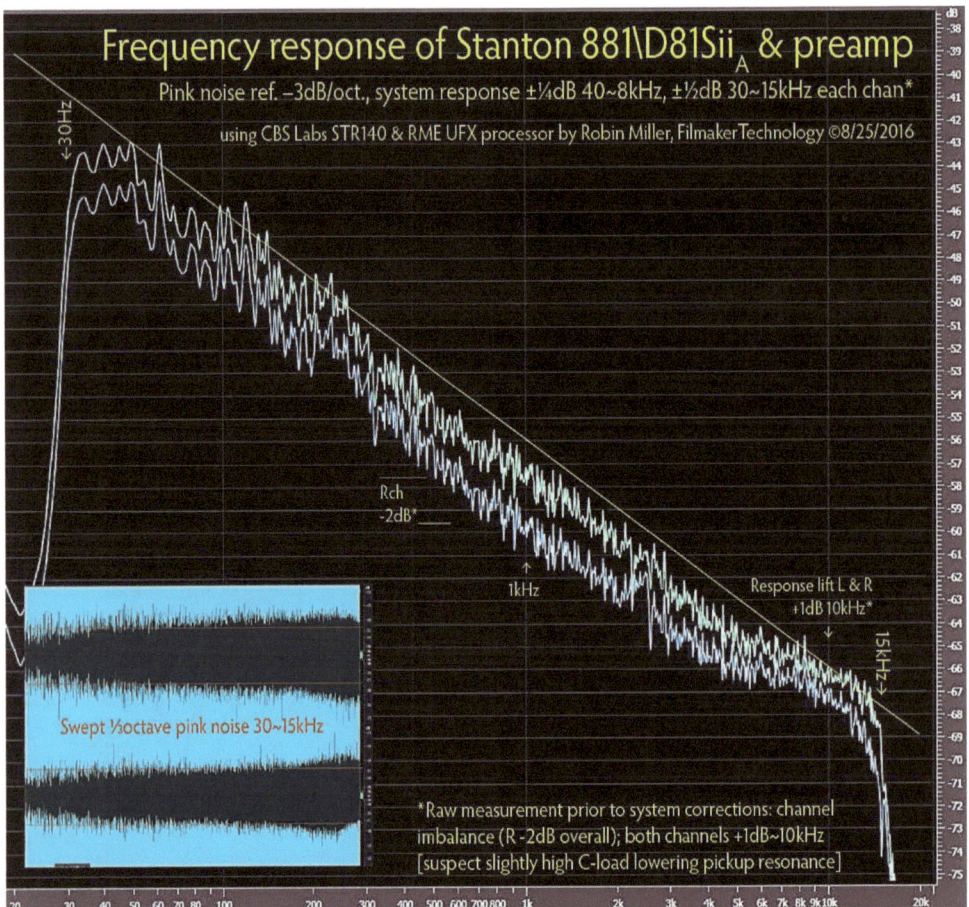

"Blue's" response incl. preamp is ±½dB over the ST-140 test disk's 30~15kHz except for intentional falloff <50Hz for rumble. A ~5% reduction in C-load would flatten the bump at 10kHz. The level discrepancy of 2dB L v. R is typical of cartridges, spelling need for preamp balance controls, though missing in most.

The spectrum analysis above shows both channels of D81Sii Blue playing a slow sweep of bandpass-filtered pink noise. Pink noise is equal energy per octave, so a flat response is represented by the straight yellow line sloping −3dB per octave. A frequency response that parallels this slope is flat. Blue's, including the preamp, is within ±½dB over 30~15kHz with two exceptions: The tilt below 50Hz is intentional in the preamp to reduce turntable *rumble;* the bump at 10kHz is adjustable by a lower *C-load,* discussed later. The 2dB discrepancy in level channel-to-channel sensitivity is typical of many cartridges although Stanton's D81Sii specification is 1dB. Easily compensated with the gain controls needed in any phono preamp for channel balance; although missing in most, it is investigated later.

NOS "Blue" had a 7° azimuth error in its tip orientation (aim), the limit of manufacturing tolerance as measured. Older styli sometimes exhibit cantilever rotation after elastomer aging. In manufacturing, the most skilled craftsmen were assigned to high-end line contact styli. While assembling the stylus, they *worried* the cantilever, rotating it in the mounting tube like a piano tuner slightly over-stretches a string, then backs off to a final position. The elastomer stores stresses from this twisting which worrying relieves. In time, further rotation from residual stresses are corrected by a gently applied fingernail. [Steinfeld 2010~13, a harpsichord tuner who interviewed Stanton employees for his Handbook.]

A robust line-contact stylus for broadcasters, DJs, & archivists

Advanced line-contacts ("parabolics") are analyzed first here because they are highest performing and longest wearing. Stereohedrons were recognized by white-gloved classical and jazz FM stations, however ruggedness was needed in a medium compliance stylus with the fine sound of the D81. Stanton's response was the D6800SL "DJ\Disco." The SL is a 0.3mil (7.5μm) Stereohedron [I] mounted on a durable, mid-compliance cantilever. Other designations were used by 3rd party sellers, but in Stanton's mostly sensible nomenclature: D is diamond; 68xx types are the company's 680 (including "calibrated" 681) series of MI pickups. *A* is for spherical, *E* for elliptical, *L* for heavy duty, and *S* for Stereohedron.

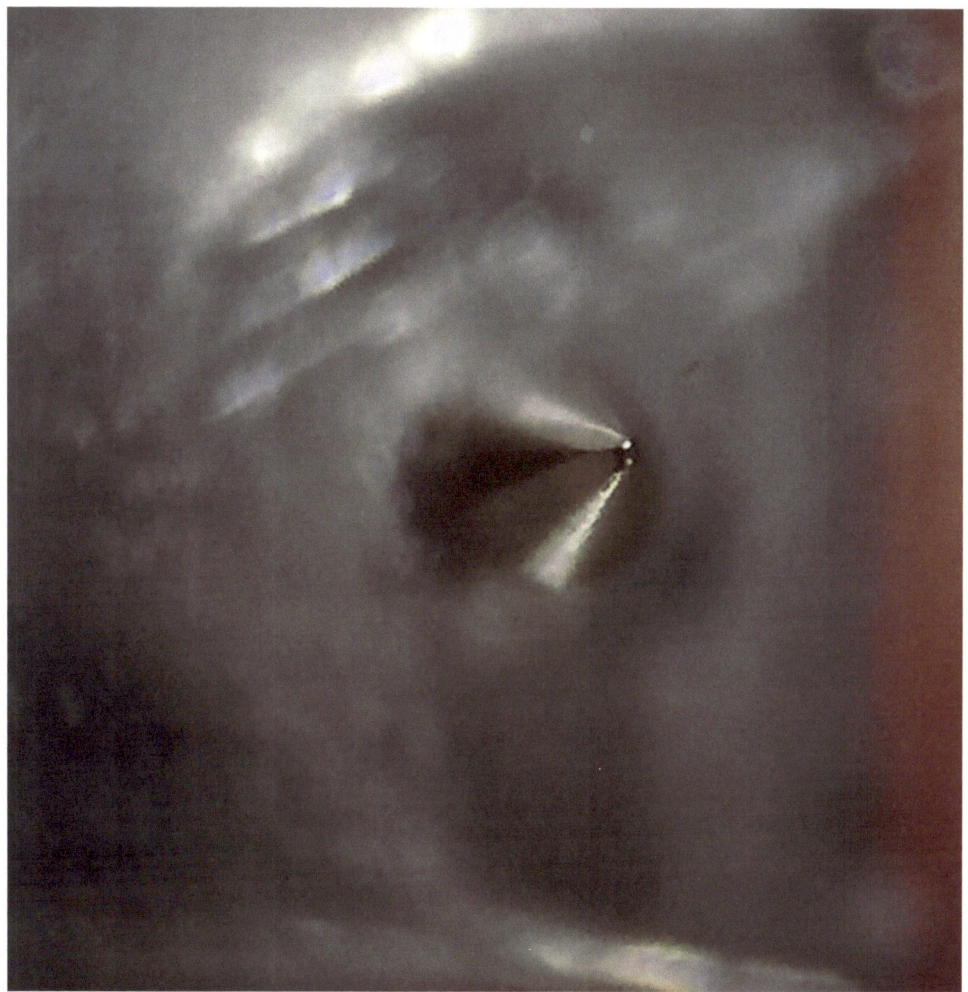

Groove approach side of a D6800SL "DJ\Disco" Stereohedron in a heavy-duty cantilever (end faces right). Tracing sides at the very tip are at the brightest bottom reflection, and just right of the brightest top reflection.

No longer made and very scarce NOS, the D6800SL exhibits high audio performance, and skip-free tracking at 2~5g VTF. At a still safe 2¾g, it is free of groove hopping and dynamic skating, anathema for broadcasters and archivists. And free of spurious distortions from any loss of contact with groove walls. Like other line contact styli, the tip is a "nude" diamond bonded through a hole in the cantilever, not on a separate pedestal. Less tip mass (moment of inertia) means highest frequencies\transients are traced more accurately.

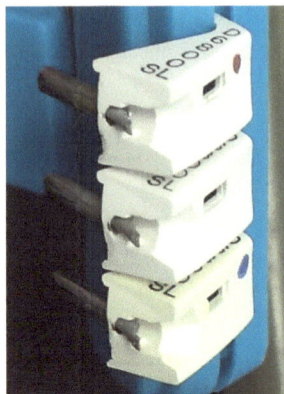 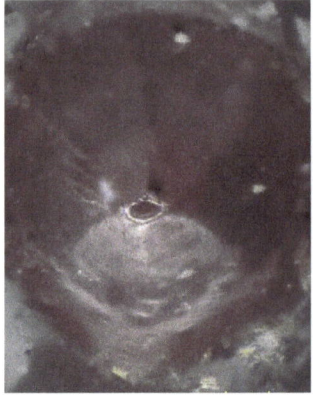 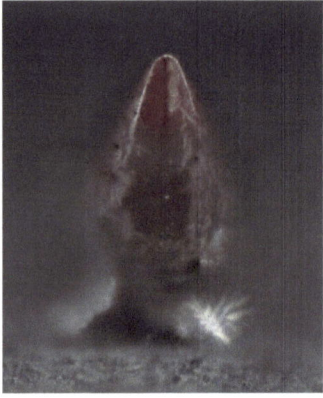

Above L: three D6800SL 0.7 x 0.3 x 2.8mil medium 12CU parabolic tip that play without fail at 2¾g vertical tracking force (VTF). **Middle:** 4th specimen from a jeweler's pink diamond dust; all four facets apparent. **Right:** The 4th pink nude's highly polished tip from front. **Below:** a near perfect D6800SL that gets the ●.

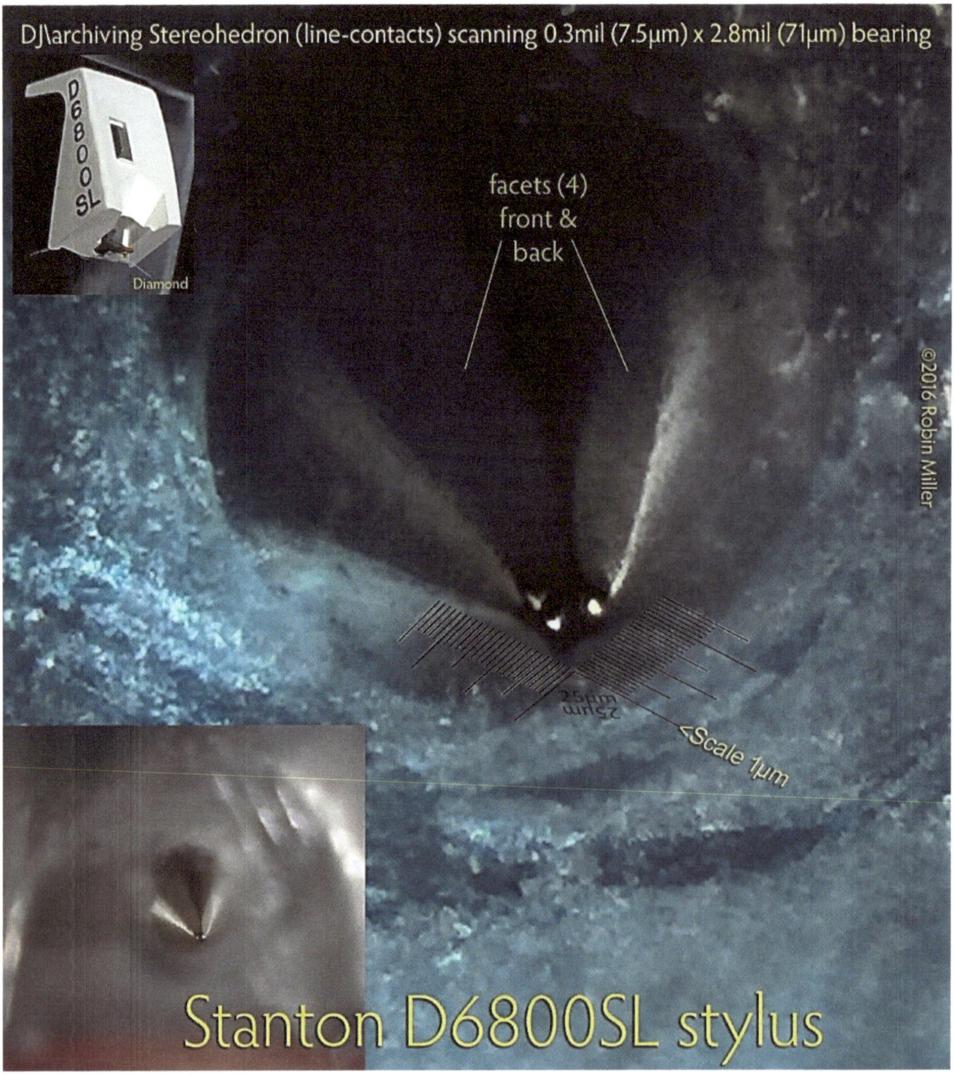

DJ\archiving Stereohedron (line-contacts) scanning 0.3mil (7.5µm) x 2.8mil (71µm) bearing

facets (4) front & back

©2016 Robin Miller

25µm

<Scale 1µm

Stanton D6800SL stylus

The more accurate audio devices become, the more neutral-performing they are, and the more they sound alike. Also with microphone and speaker transducers, neutrality is desirable for archival "ripping" of pristine vinyl, including digital *ingestion* ("transfer"). Wear is of less concern as a pristine disk might not need playing more than very few times.

Neutrality (low distortion, flat response) has been the goal of most enthusiasts from the dawn of the hi-fi hobby ~1950 until digital audio. And the objective of audio research and record mastering from the beginning. But even with accurate replay and listening space acoustics that would reproduce nuanced *real* sounds, "audiophiles" since claim differing and changeable preferences for coloration. As taste matures, all seek line-contact styli.

Between line-contacts and sphericals in quality – the elliptical stylus

The examples in this paper are mostly offerings by Pickering and renamed Stanton, New York, who in that incarnation made cost-effective cartridges & styli for consumers and professionals, with crossovers among their 109 choices of needles. This limitation reflects the author's choice to focus limited resources on a fine yet practical line of phono pickups with interchangeable styli. However, the principles explored apply to many other makers, most of them offering middle-grade elliptical styli similar to those in this section.

From the phonograph's invention of reproducible sound in 1877 to well into the 20th century, all replay styli were hemi-spherical, making circular contact of the walls of the groove. Written for hobbyists and professionals today, this book makes no secret in not advocating use of spherical stylus tips (sometimes called in error "conical.") [9] Except for riding over scratches and clumps in the dirtiest disks, sphericals today only apply to the lowest quality phonographs. From the introduction of the microgroove high-fidelity (mono) recording in 1948, spherical styli continued to produce distortion, cutoff high level high frequencies (HF) in inner grooves, and in a few plays can "erase" HF by "re-cutting" the sharp turns they cannot negotiate.[10] Literally cutting corners! Later we'll explore how sphericals are still useful for certain LPs and coarse-groove SP 78s.

For an early grasp of what a stylus can be, we've begun by exploring high-performance line-contact reproducers. Today these are scarce and expensive. So next are middle-range elliptical styli in three basic designs. And offered in dozens more variants and model designations. By the sharpness of their sides (scanning or tracing "radius"), they are 0.4, 0.3, or 0.2mil (10, 7.5, or 5μm). Compliance Units span 10~20 "CU." [11] The fattest 0.4mil (10μm) is a hardy tracker, and has better HF tracing and distortion compared to a spherical of 0.7mil (18μm). Comparative wear and cutoff frequencies of all shapes are on p22.

The sharpest elliptical, with a cross-section side radius of 0.2mil (5μm) implies best HF response, lowest distortion, but at the cost of highest wear, offset by lower VTF and arm mass, and higher compliance (loosest springiness) of the cantilever. A compromise 0.3mil elliptical is the one-size-fits-all offering by new and replacement makers today.

NOS or new, *caveat emptor* choosing a stylus. "Pre-owned" at auction should be certified under a microscope. The data and calculations from p22 have been simplified in the next table, generally rated 1~5 in sound quality by stylus shape. Then examples of ellipticals' construction and wear are illustrated in the micro-photo posters that follow.

---

[9] All styli begin as conical and are only conical well above where they contact the groove, where shape is irrelevant. The more precise terminology in chronological order is "spherical," "elliptical," and "line-contact."

[10] Per p22, for 0.7mil (18μm) sphericals >1958, worst case f_3 =3.4kHz; 5.8kHz for 0.3x0.7mil ellipticals today.

[11] Compliance Units in mN/μm, the cantilever's *springiness* when 1 milli-Newton of force causes 1μm deflection.

Phono styli close-up – how they work; how they wear

General stylus quality rating by tip shape ©2017 Robin Miller, FilmakerTechnology v8/24/17

Rating	Tip shape	VHF tracing	HF distortion	Wear*	Notes
5	Line contact	Very good	Lowest	Lowest	Critical VTA\SRA by user
4	0.2mil hyperelliptical	Very good	Lowest	High	Critical mfgr aiming of tip
3	0.3mil middle elliptic.	Good	Moderate	Mod	Requires mfgr aiming $\leq \pm 7º$
2	0.4mil common ellip.	Fair	Mod~high	Low	Forgiving mfgr tip aiming
1	0.7mil spheric (conic)	Poor	Highest**	Low***	Non-crit VTA\SRA by user

* at same VTF; for typical VTF by tip shape see table "Phono record groove & stylus wear & HF cutoff by tip shape" p23.
**Most prone to "sibilance" distortion of high level HF inner groove.   ***plus HF "erasure" of high level inner groove.

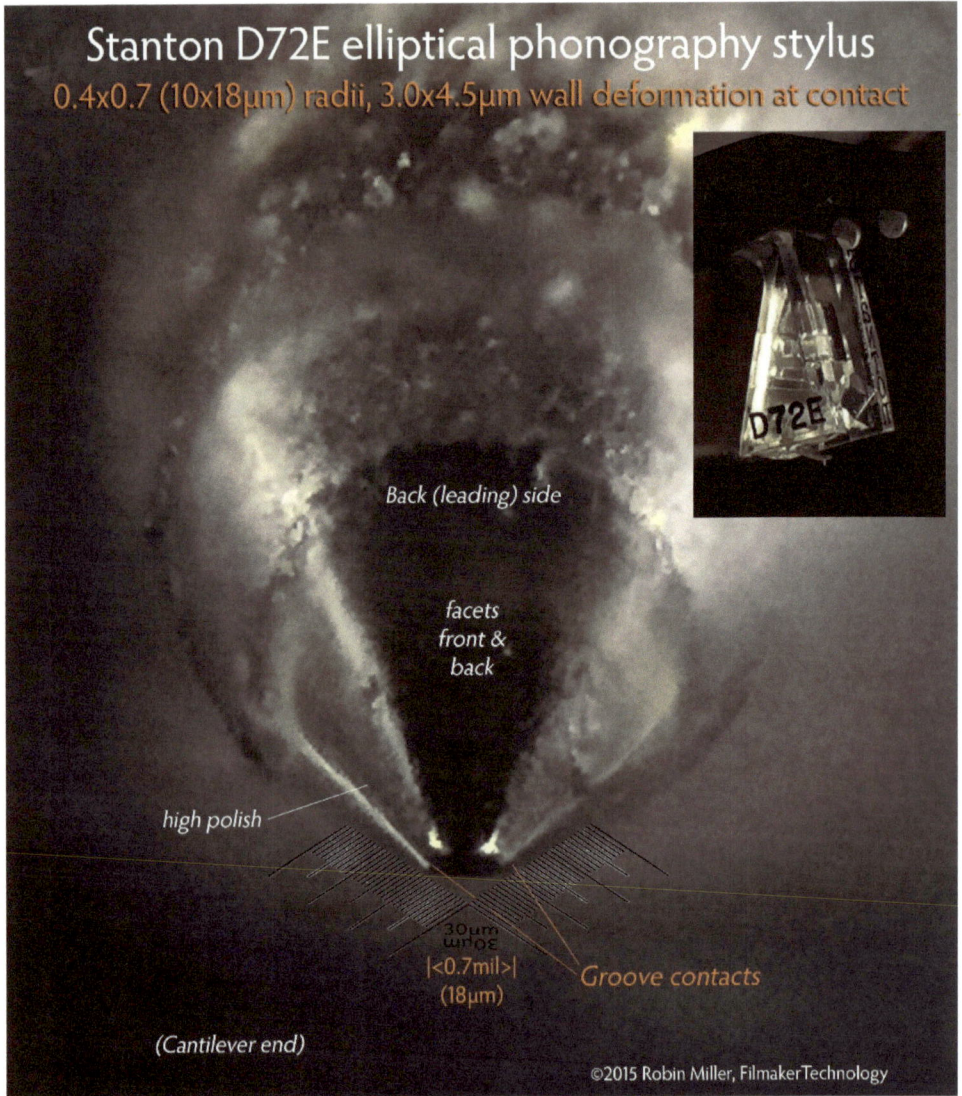

An unworn 0.4x0.7 elliptical, precisely ground, heavily truncated, and dead-ahead aim.

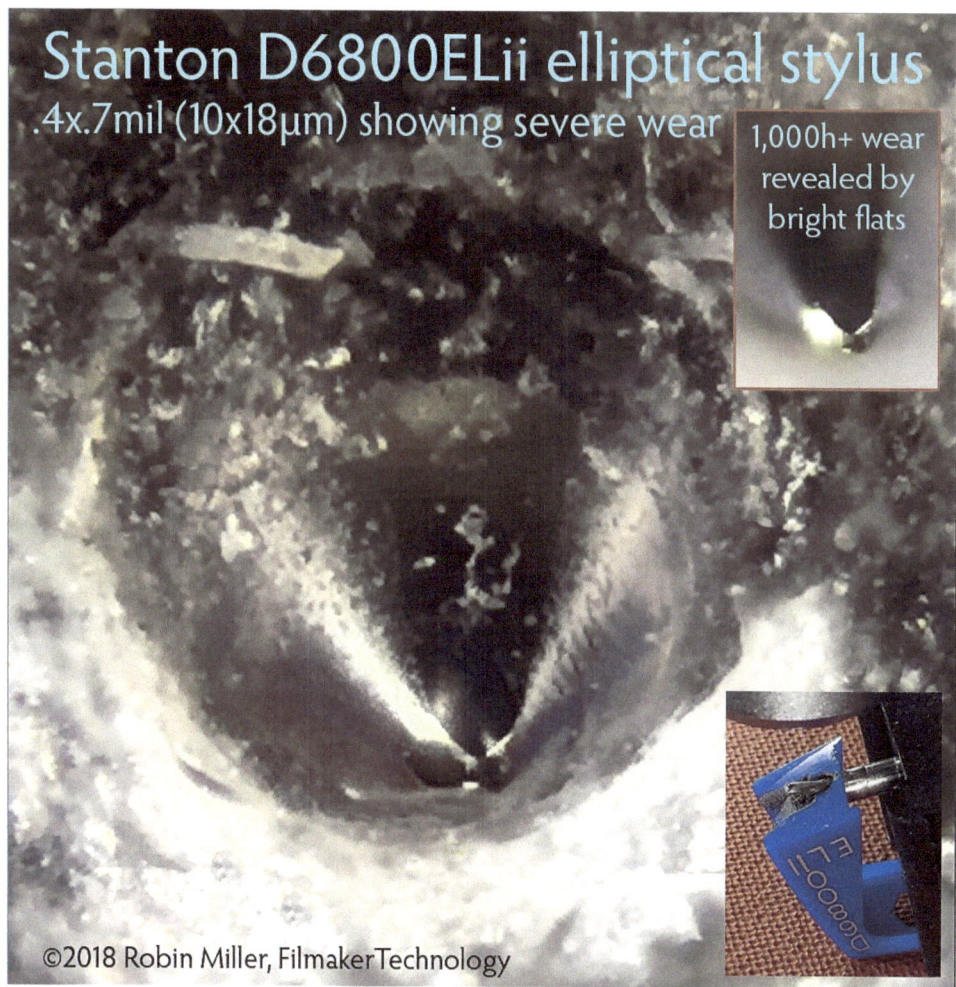

Stanton D6800ELii elliptical stylus .4x.7mil (10x18µm) showing severe wear

1,000h+ wear revealed by bright flats

©2018 Robin Miller, FilmakerTechnology

Submitted by a radio station client for certification (and obvious need for replacement), an over-worn (>>1,000h) 0.4x0.7 elliptical has severe flats. Both sharpened perimeters undoubtedly have damaged permanently many records. The far broader flat that rode the outside wall is from too much anti-skating.

Unlike the circular contacts for spherical styli, for narrow-sided line-contacts' and ellipticals' the elongated in *race-tracks* up & down make it critical to parallel the groove wall's high frequency waves. This is set by the tonearm's Vertical Tracking Angle (VTA, aka Stylus Rake Angle SRA) standardized at 15°. Ellipticals for moving iron (MI) cartridge are represented by those for Pickering XV15 and identical Stanton 680 (and calibrated 681), ranging from low-mid-end 0.4mil D6800E and DJ-duty -EL to narrower-sided 0.3 -EE & 0.2mil -EEE models. Or moving magnet (MM) cartridges represented by professional 880 (and calibrated 881) or Pickering's popular V15 "broadcast standard" and identical Stanton 500, with ellipticals designated 0.4mil D51xxE and DJ-duty -EL, to narrower 0.3mil -EE.

Again, the examples in the posters apply to nearly every manufacturer's styli that are more readily obtainable, as long as their tip dimensions are published. Some ellipticals examined are severely truncated at the tip's end for avoiding dirt at the very bottom of the groove of consumer disks that are seldom cleaned. Exceptions are evident in the posters.

Phono styli close-up – how they work; how they wear

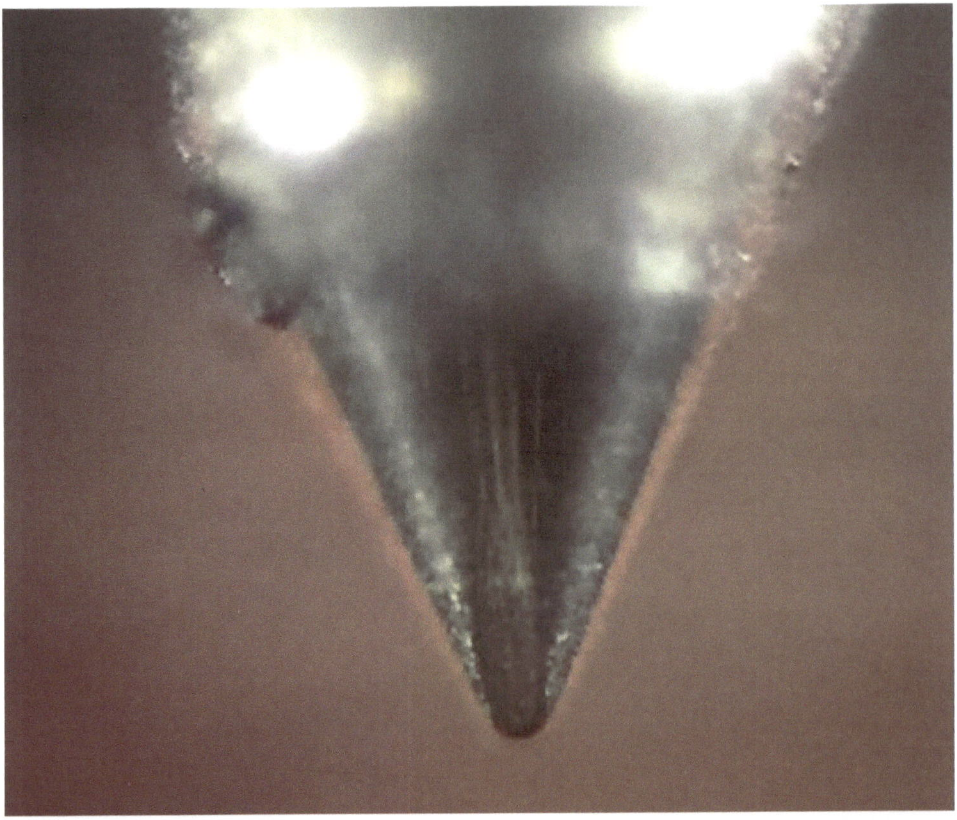

A somewhat rotated (but acceptable) D5100E 0.4x0.7 from the back (groove approach) ***at top*** and from the spindle side at ***bottom L & R***. This fattest elliptical is safest in "semi-automatic" turntables, that return to rest following a record side, or "fully-automatic" that "cue" from rest to play from the beginning. More robust -EL versions on heavier duty cantilevers are also those recommended ellipticals recommended for "DJ" use.

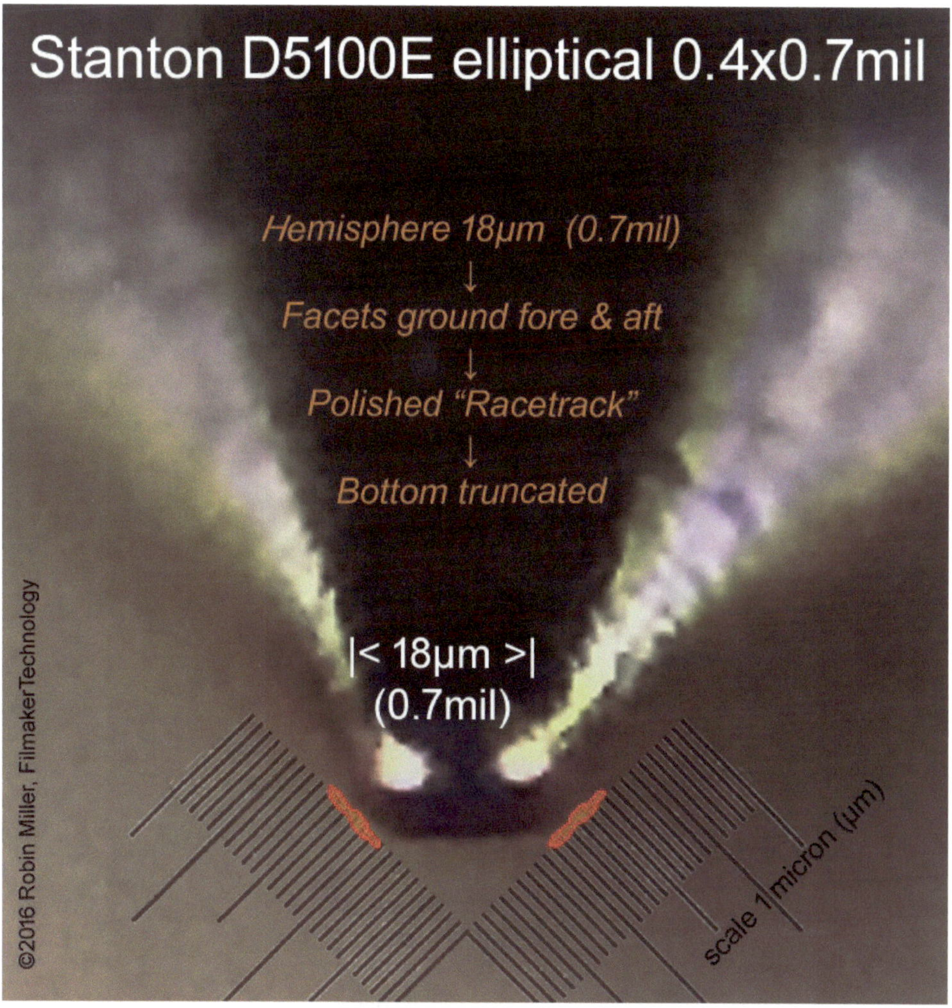

Stanton D5100E elliptical 0.4x0.7mil

Hemisphere 18μm (0.7mil)
↓
Facets ground fore & aft
↓
Polished "Racetrack"
↓
Bottom truncated

|< 18μm >|
(0.7mil)

scale 1 micron (μm)

©2016 Robin Miller, FilmakerTechnology

All tips are rough-ground to a cone, then a hemisphere, then are ground two facets for an elliptical, or four for a line-contact. To make this popular (low\mid $) elliptical: off-vertical facets are ground flat fore & aft, polished, and severely truncated to avoid flotsam fallen into the uncleaned groove.

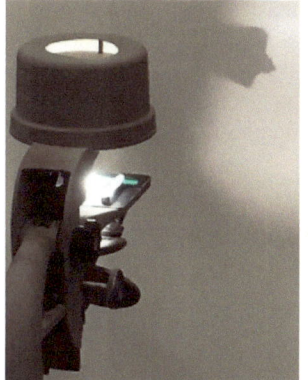

For high volume QC, an 80x shadowgraph was easier on workers eyes than a microscope. in the middle image, this "fat elliptical" is less than doubly wider from the back, than from the side, at right.

Phono styli close-up – how they work; how they wear

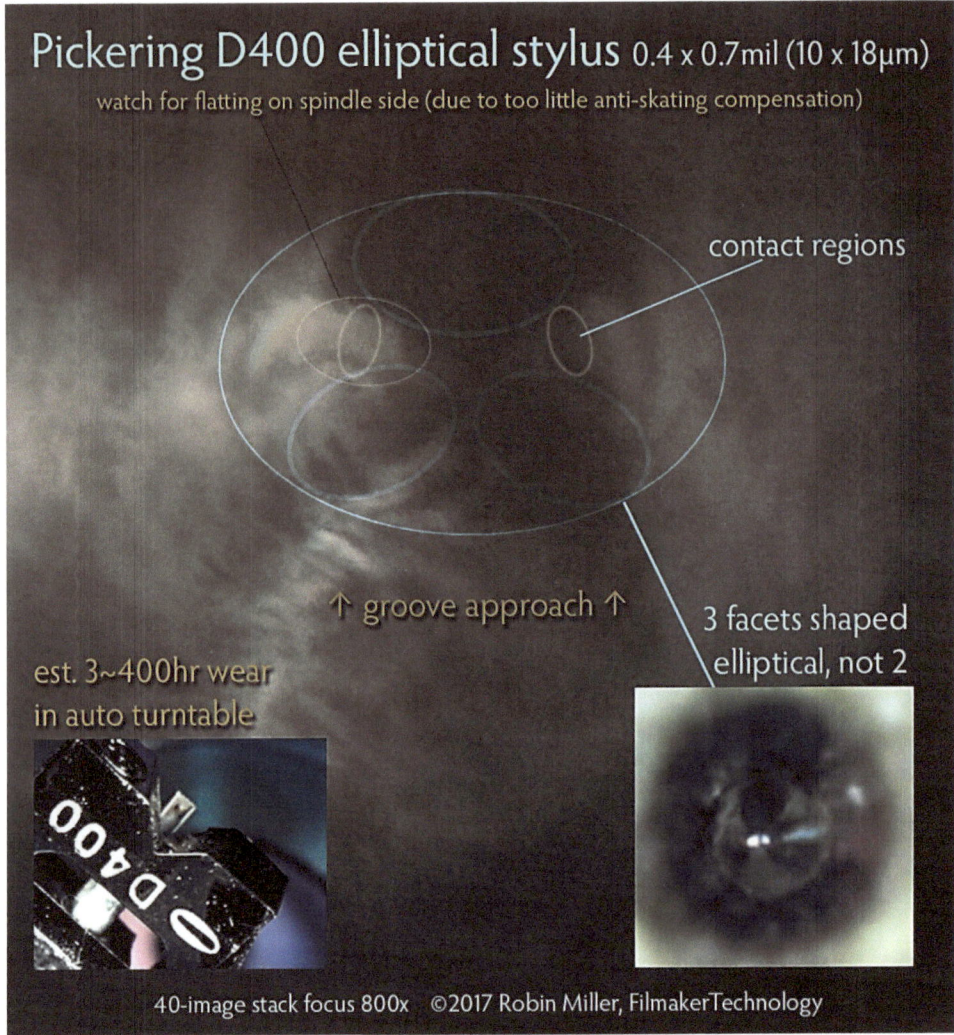

At the optical microscope limit of ~800x, a D400 adds a third flat, so tracing is aft of the groove's passing.

A Pickering D400 0.4 x 0.7mil "fat elliptical" above was a favorite for audiophile automatic turntables made by Dual, in a XV-15 cartridge. This mildest elliptical shape nearly equals the pressure and wear of the spherical at the same VTF, but with somewhat improved HF response and lower erasure by the inner groove.[12] Regarding distortion, the D400 has *three* flats ground, moving its racetrack-like contact patches to trail in the groove. This shape's off-perpendicular tracing may have been intentional coloration, as explained…

If not at the point of perpendicularity to the groove's undulations, a D400 adds a form of distortion ("*poid*-like," to be explored later). The bright coloration of all-order harmonics contributes to this stylus' popularity with less than discerning listeners with middle-grade automatic turntables. Later, narrower ellipticals with perpendicular side scanning radii of 0.3mil (7.5µm) and 0.2mil (5µm), which for comparable wear necessitated tracking at lower forces in lighter arms, elevated their critical vertical tracking angles (VTA/SRA).

---

[12] Relative wear at intermediate VTFs for any stylus can be interpolated using the spreadsheet on **p22**.

To simplify implementation of phonograph systems by users, manufactures agreed to a "Universal P-mount" with fixed alignment (non-user-optimizable). In either MI or MM cartridge variants, the agreement called for a standardized changeable plug-in form factor, 6g pickup weight, and VTF of 1½g. Despite different model designations, styli were interchangeable with those for regular pickups that have ears for mounting in a head-shell using screws on ½in (12.7mm) centers. The D6800<u>E</u> was also a D72<u>E</u> (indicates 0.4mil), but in a clear grip (**overleaf**). Pickering and Stanton had half a dozen P-mount offerings each, mostly ellipticals, top models Stereohedron, but only one lowest model a spherical.

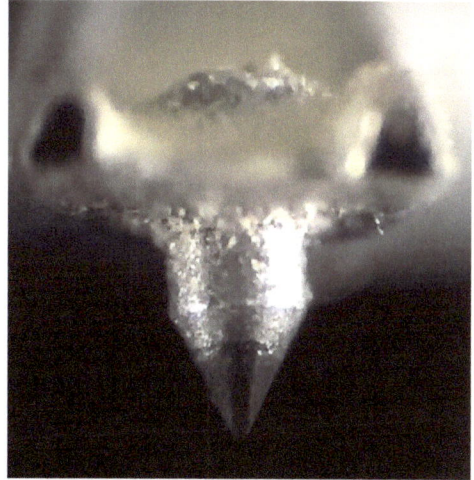

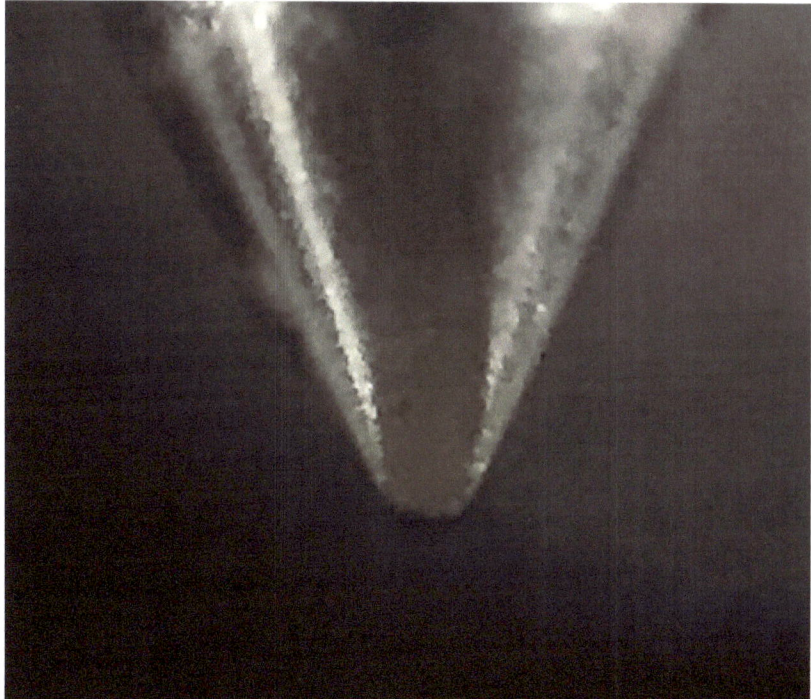

"Universal P-mount" for reduced fiddling by consumers standardized plug-in cartridges, tracking at 1½g, and with a range of stylus quality. The D72E is a 0.4 x 0.7mil elliptical polished from two ground facets fore & aft. Does the slightly canted truncation impact the groove?

Phono styli close-up – how they work; how they wear

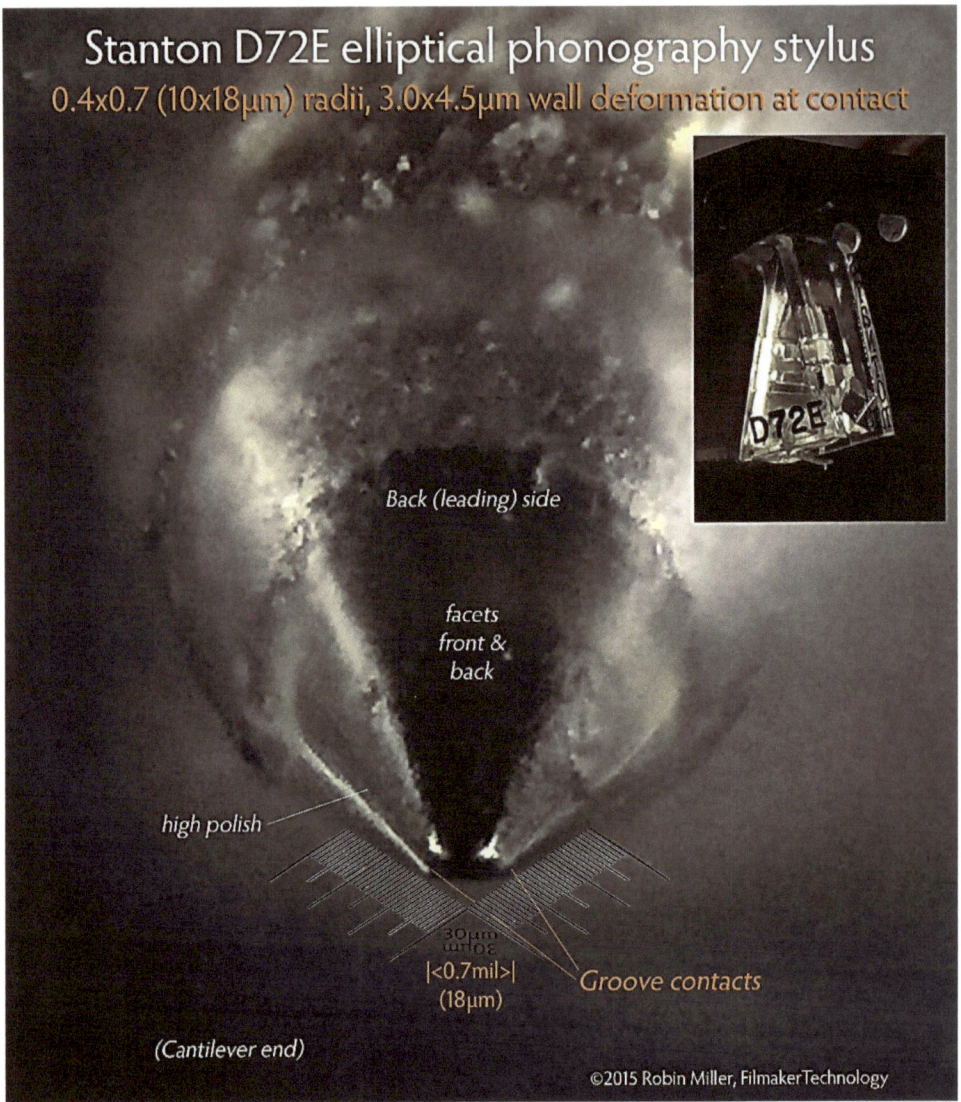

A desirable size, and more available today, is the 0.3mil elliptical.[13] **Opposite** at high magnification is an "EE" after 2~300hr of service when as a teen my son DJ'd school dances. Before imprinting designations on grips, it was identified by the color of a raised box of gold for the 0.3mil EE. It was Stanton's the best performer for V15\500 bodies, specified for a VTF of 1~2g (no line-contacts were available). Within the red ovals are very low and nearly symmetrical patches not yet worn flat. However even with my son's reasonably careful use (and my attending to its alignment), the same model stylus in his second turntable might have been stressed, its diamond came unglued and gone forever between floorboards! A less costly 0.4mil red box D5100E, or -EL "heavy duty 0.4mil," might have been more appropriate for non-critical DJing (wedding receptions, dance clubs) – possibly with a bit more glue? Today my son's older son, on his own grampa-refurbished turntable, has his dad's remaining D5100EE, with 7~800hr remaining if cared for.

---

[13] 7.5x18μm, tracing radius x contacts width, and wide VTF range 2~5g. Expert Stylus UK recommends ≤3g.

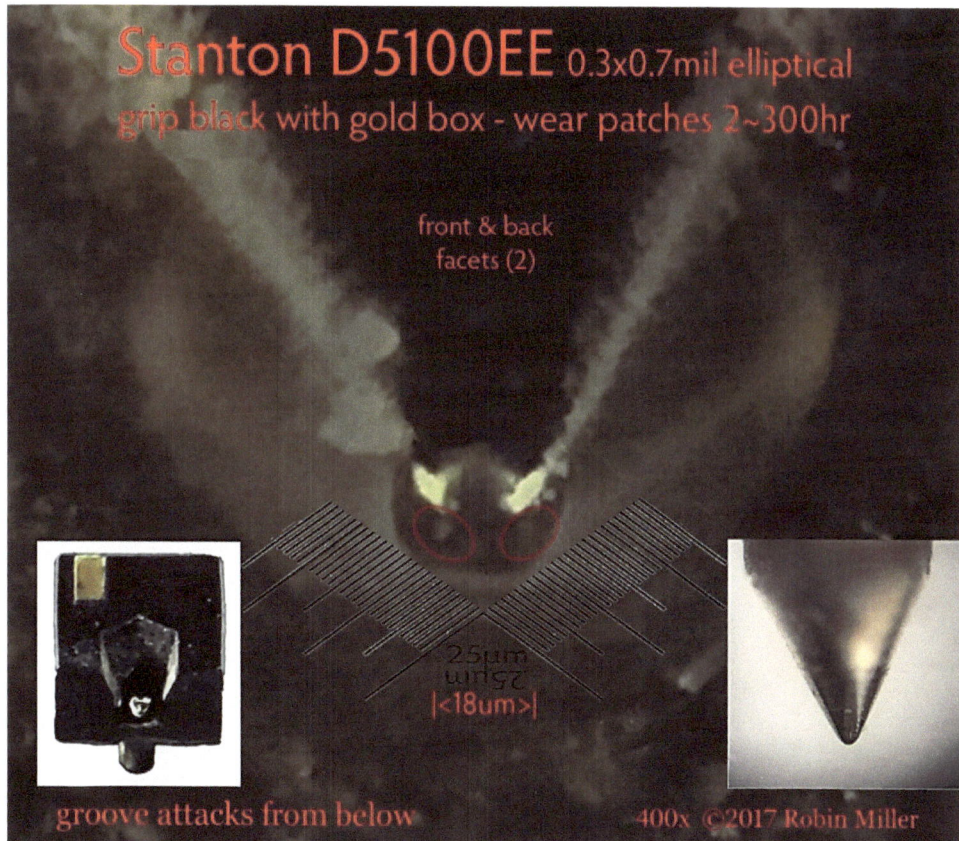

Stanton D5100EE 0.3x0.7mil elliptical
grip black with gold box - wear patches 2~300hr
front & back facets (2)
25μm
|<18um>|
groove attacks from below
400x ©2017 Robin Miller

Today the most available among reasonably good functioning styli that makers and dealers offer as new units and replacements are the middle-of-the-road 0.3mil (7.5μm) side radius tips, bonded on moderately compliant cantilevers, tracking at 1.5~5g. At a safe 2g VTF, a 0.3mil elliptical has 70% higher high frequency response and lower distortion and 26% more wear as a standard 0.7mil spherical, but 3~4 times as much wear as a same-sided line-contact. Yet these are about right for casual vinyl users who keep their records clean.

The narrowest elliptical, that evolved into the line-contact

Prior to the 1972 Shibata and other advanced line-contacts, the narrowest 0.2mil (5μm) ellipticals were specified to track at ~1g. Ellipticals' contacts are "racetracks," bearing up groove walls somewhat more than their horizontal scanning\tracing width. Before Stanton used the EEE designation, the MM Pickering D3001E, **next page**, is a nude that exhibits the same 8.7kHz f_3 cutoff (worst case) and low distortion as the same-sided 0.2mil D81Sii Stereohedron line contact. And with its shorter racetrack, the elliptical is less sensitive to vertical tracking\stylus rake angle (VTA\SRA). But it presses heavily to cause high wear.

Under the passing stylus' extreme pressure, plastic vinyl instantly rises in temperature, the groove walls briefly deforming at the contacts. The vinyl doesn't fully melt, like ice under an ice skate to glide on a thin film of liquid. The vinyl softens and is indented at the instant of play, then re-solidifies a few seconds after the stylus passes, before possible onset of permanent damage, such as immediately repeated scratching or "cue burn" by a DJ. As said, reducing wear proportional to the inverse of higher contact area led to the line-contact.

Phono styli close-up – how they work; how they wear

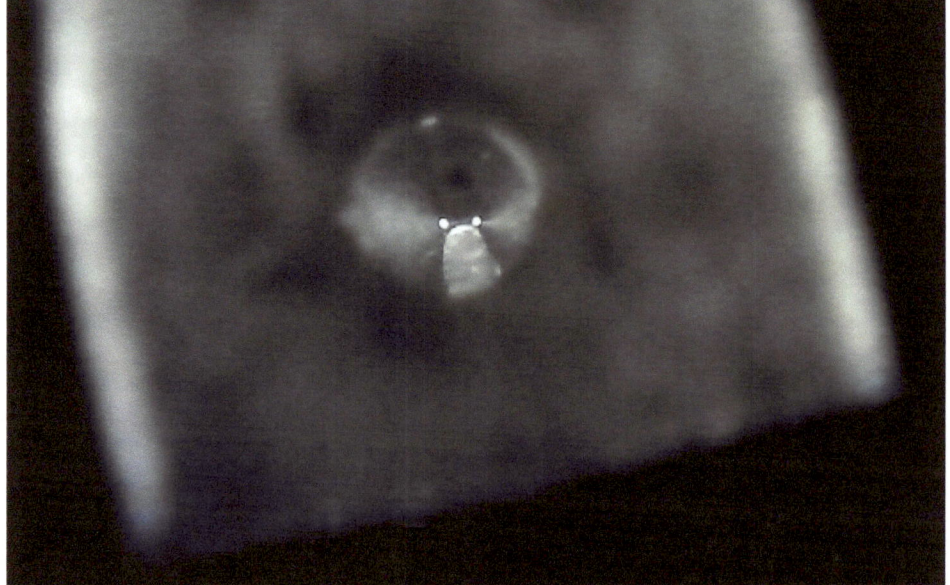

D3001E is fine elliptical of 0.2 x 0.7mil nude gem chip. Compared to a spherical's round contact circles, the contact areas are "racetracks" up\down the groove walls, so SRA\VTA is the second most critical. Even tracking at 1g VTF, the D3001E's relatively small contact area accelerates wear of both stylus and records. The flats (not the pinpointy highlights) indicate precise aiming.

The lowly spherical (sometimes called "conical") needle

Why spend as much ink & paper over these stylus tips? Because they dramatically affect the sound. Rendering what's in the groove is not easy. It's subject to all manner of issues, physical and mechanical. Gyrating wildly up to 20,000 times a second for up to 1,000 hours (1,000 LPs both sides), it's a wonder even the simplest sound as good as they do*!*

Easiest (cheapest) to manufacture are spherical styli, less precisely called "conical." All tips begin as conical, tapering to groove width, but are finished in the three main profiles, as illustrated on **p20**. Sphericals are simply that basic cone, ground at the tip to a half ball. With their circular contact patches, the SRA\VTA of sphericals is nearly irrelevant except for maximizing output signals by the cantilever's magnet or iron slug orbit amid the coils.

But for many recordings, sphericals create the most inner groove distortion, termed imprecisely by audiophiles as "sibilance," named for the raspy sounding effect on vocal sibilants. For more money, and greater user\installer knowledge, an elliptical or line-contact stylus will result in far less occurrence of sibilance (Later we will dive even deeper into what causes the distortions due to stylus shapes not always fitting the groove.)

In some cases, a simple spherical tip is an OK choice. For example, softer ballads by voice and piano contain far lower-level high frequency energy, for which the spherical shape might fit the groove just fine. Such music ranges from Alicia Keys and John Legend to Schubert Lieder. With energy spectra falling gently at high frequencies, the spherical or fat 0.4mil elliptical shapes can negotiate the gentler turns in the groove. The same applies to older recordings with limited bandwidth. Meanwhile, cleaning of disks becomes more tolerant, as the broad contact areas glance more quietly over dirt and scratches, instead of playing every snap, crackle, and pop, as may sharper elliptical and line contract profiles.

Below is the Gibson-Stanton company's ambiguously designated N500-6, a descendant of a D5100A(L?), in a ferret-faced grip. Examined closely below, there are no facets, so it's a low-end spherical, though not shyly priced, its cantilever rugged for DJ scratching. If "-6" means a reduced diameter of 0.6mil (15μm) – Gibson-Stanton don't say – then its f_3 by the inner groove is 3.7kHz, a scant 9% above more common 0.7mil (18μm) types, and it generates 9% less distortion, aka "sibilance." But wear is higher by 60%, as in the table on **p22**. Is this performance v. wear by reducing the tip from 0.7 to 0.6mil a good trade?

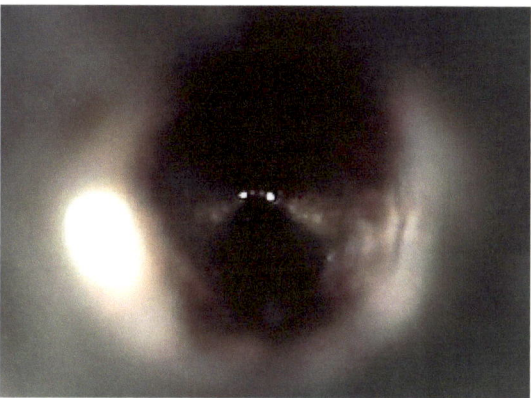

A N500-6 spherical 0.6mil (?) suitable for Stanton's workhorse moving magnet 500 (originally Pickering V15). Updated as a "Mark II," with stronger Samarium Cobalt (SmCo) magnet, its resonance is raised from 13kHz to 15kHz. No longer in new manufacture, a large remaining stock of both is sold under Stanton owner Gibson.

Having reached the bottom of the stylus pecking order, readers will have learned which to seek next. Again, it is the most critical component in the phonograph system affecting (distorting) the sound. Well-made digital recordings will sound *better* re distortion unless *clean* is not a listener's preference. Prior to playing the groove, audiophiles' wailing over use of digital recording to source, mix, or intermediate in mastering new vinyl is nonsense. Only making the vinyl pressing less distorted, it is no more "impure" than using tape in pre-pressing steps. Even worse technically, cassettes and cylinders still contain music to enjoy.

Before microgroove, 60+yr of SP 78s & electrical transcriptions (ETs)

Before the micro-groove LP or 45 (introduced in 1948 and '49), the phonograph gauge was "wide-groove" ("coarse-groove," or standard play, SP) 2~8mil (50~200um) wide, and with wide needles to fit. Most of the 140+ years of recorded releases require SP needles! Initially of steel for playing a single side in acoustic players, styli evolved to use industrial sapphires, and then diamonds in two shapes: spherical for consumers; elliptical shapes for conservators. We'll explore all three. ***Do not play disks with the wrong needle!*** Playing 3mil 78rpm disks with a ≤1mil microgroove stylus *reduces stylus life to only 2~300 hours*, same as for SP styli, and has gouged many disks brought to the author to restore digitally. Playing microgroove disks with an SP stylus just skates across the disk and scratches it.

As a compromise width between 2.0 for ETs and 3.0 for most 78s, next are 2.7mil (68μm) spherical nude diamond tips. For restorations, the author wet-cleans shellac 78s, or the SP tip quickly collects dirt, as shown **below** right. Easily clean needles by dabbing in *MagicEraser*. *Stylus life is 2~300 hours*, ⅓ that of an LP stylus due to 1) high abrasion of "shellacs" containing ground slate; 2) no deformation of the hard walls means the contact area is smaller, so wearing pressure is higher; and 3) 12in 78s run 2.35 times the linear groove speed of LPs. One may not notice, as a 78s running time is only 3~4½min.

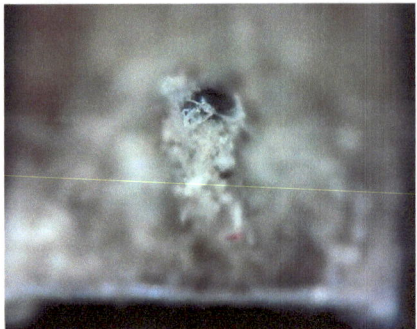

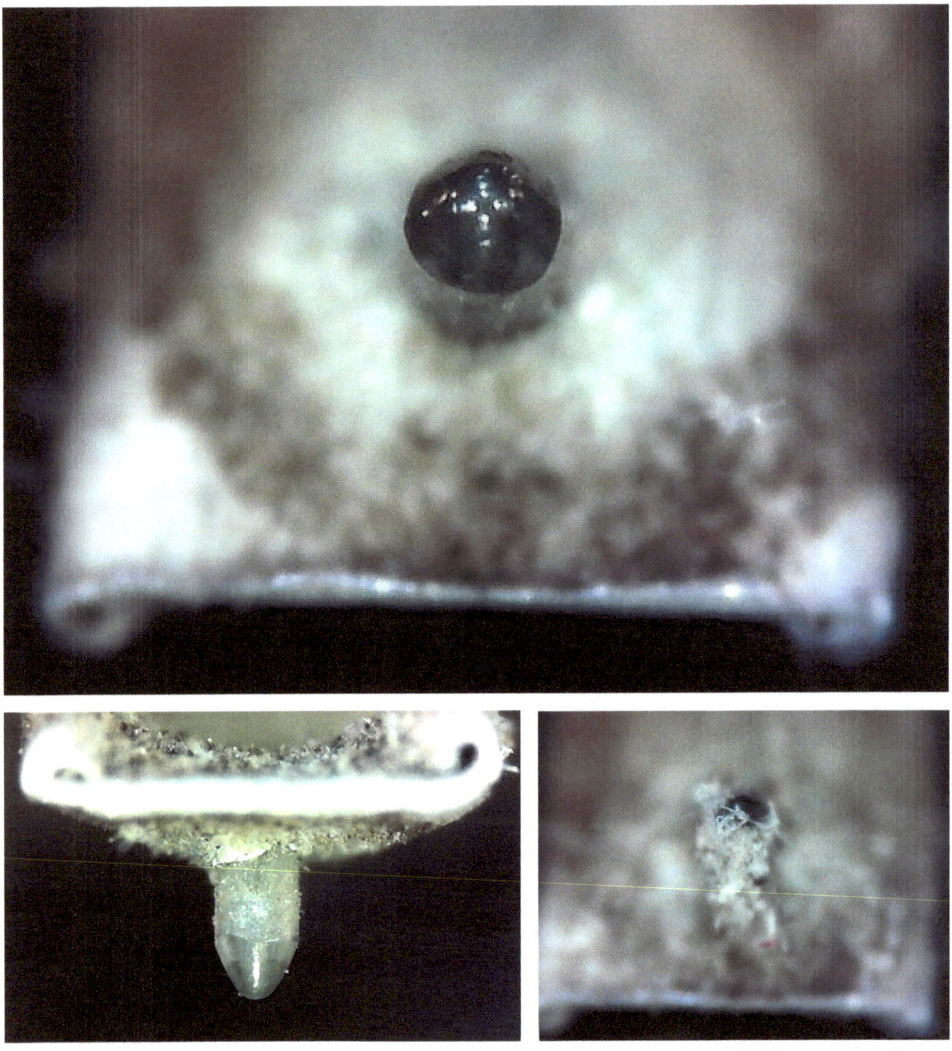

For wide groove standard play (SP) 78rpm and 2mil electrical transcriptions (ETs), a compromise 2.7mil tip for both consumers and broadcasters. ***Top & left,*** a nude gem of the Pickering DCF4527 for moving iron (MI) XV-15 and 680\681 cartridges, identical to the Stanton D6827 shown in the poster **opposite**. ***R,*** dirt accumulated playing a filthy 78 before a few quick dabs in *MagicEraser*.

Stanton D6827 "nude" 2.7mil (69μm) standard (coarse) groove MI stylus

© 2017 Robin Miller, Filmaker Technology

SP (wide\coarse-groove) restoration\archiving styli

Wide-groove (SP) widely ranges from 2mil (50μm) for ETs to 8mil (200μm) for Pathé vertical shellacs running ~80rpm. For archiving work, the restorer interchanges styli of varying dimensions until s\he finds by ear the groove wall depth with the least damage. The next poster shows one of a set intended for conservators of most 78rpm SP and radio ETs, a 0.5x3.0mil (12.5 x 75μm) elliptical, interchangeable in a workhorse MM Pickering V15 or Stanton 500 cartridges. This D5130EJ lies in the middle of the range 0.4 x 2.0 mil (10 x 50μm) for pristine radio's electrical transcriptions (ETs), to larger for worn 78s.

The D5130EJ poster overleaf illustrates the narrower contact areas than a spherical tip. As with microgroove LPs & 45s, elliptical profiles increase wear of both groove and tip. Smaller 2.7mil spherical SP needles are a bit gentler. The narrower scanning sides of the D5130EJ traces high frequencies with lower pinch-effect ("sibilance") distortion. Finer tracing is needed less for older SP recordings, with limited HF baked in, and for the 78's linear speed.[14] Over the decades, the high frequency limit of 78s doubled from 5kHz to 10.

---

[14]Linear speed (in/s) =60πdr, where d =diameter of the groove; r ranges by format between 16⅔~100+rpm.

Before microgroove, 60+yr of SP 78s & electrical transcriptions (ETs)

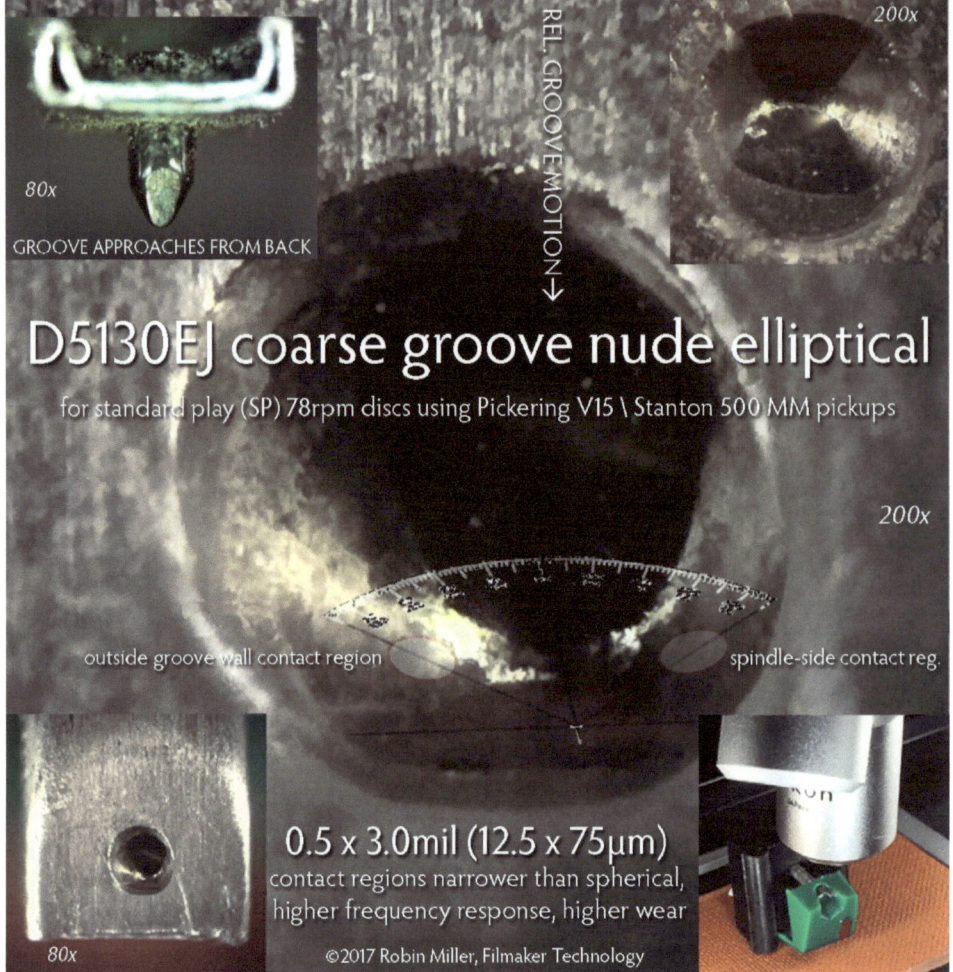

Many cartridges & styli were made in "vinyl's" heyday

Not including mass-marketed crystal and ceramic\piezo cartridges, others' pioneering electro-magnetic phonograph "translating devices" are deserving of their own posters, papers, and books. Early (1928), Western Electric (aka Westrex, ERPI) made moving coil pickups for talkies and Radio. The stereo MM pickup may be attributed to Electroacustic GmbH (ELAC, German patent 1957), cross-licensed from 1962 to Shure & RCA.

Other key developers were Bachman's "variable reluctance" (VR) aka induced magnet (IM, fluxvalve, aka MI) while at GE. [At Columbia on Goldmark's team, it was Bachman who invented the LP]. Astatic (USA, Canada), Walco, and Electro-Voice EVG were re-packagers of OEM styli, while Pfanstiehl (Switzerland & USA) repackaged others' (3-digit models) and manufactured their own "generic" replacements (4-digit models, "4nnn").

Choosing a stylus is more of a challenge today, given the plethora of misrepresentations online. Active today making quality products are Goldring, Ortofon, Audio-Technica, Nagaoka, and others. Look for products with meaningful specs. Then your most useful guides to choosing a stylus profile are the research data and calculations tabulated on **p22** that have been simplified and generally rated by stylus shape in the abridged table on **p34**.

Generic "aftermarket" 2.7mil Shure replacement, Pfanstiehl #757-D3, also shown installed in an RCA-branded Universal Stereo broadcast pickup MI-11865 below left (OEM Shure 4-coil cf. 2-coil of ELAC) Although the first pickups, later were reintroduced the more complex and expensive *moving coil* (MC) cartridges with fixed styli, lower compliance, and higher distortion. But we.

Decca-London (UK) made a unique MI cartridge with a non-user-changeable 0.65mil (16.5μm) spherical on a Z-shape cantilever, restrained by a tiny nylon rope, tracking at 3g, and sounding very good, although risking wear. And more unique in having three coils: one detecting the lateral motion of the stylus, two for the vertical motion. The result is fine monophonic replay (shorting out the vertical coils), but its stereo soundstage depends on the difference coils matching. However its precise soundstage balance cannot be corrected simply by individual gain controls in the preamp because that would upset the contributions of the single lateral coil. They were graded by body color, Blue to Violet to Gold, also adding advanced stylus tips. Shown is a Gray, a Blue selected for export to the USA.

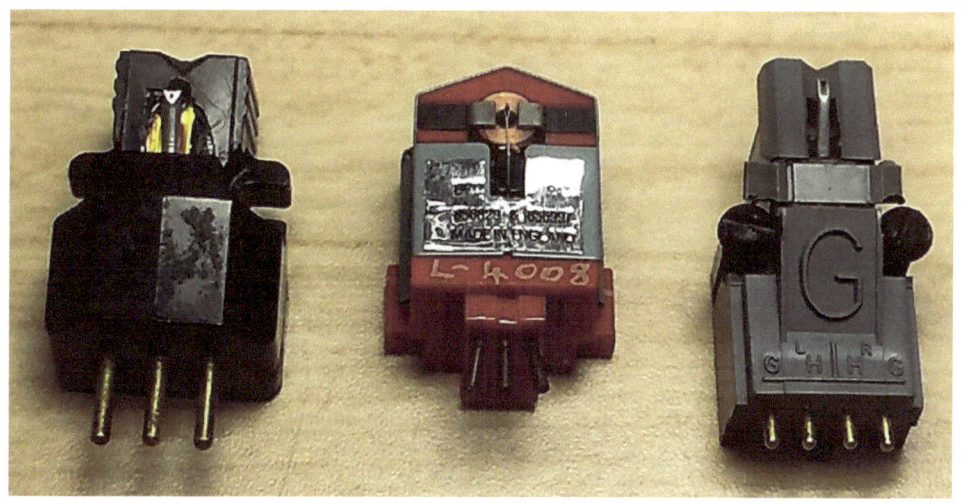

L: RCA "Universal stereo" 1963 (OEM Shure MI 4-coil M3 & M7 (with improved body that didn't bottom on warped disks). In turn did Shure obtain ElectroAcustic's (ELAC, Germany) 2-coil 1957 design for patent protection? *Middle:* Decca-London Export 0.65mil (16μm) spherical with cantilever restrained by a nylon rope, shown close-up **overleaf**. *R:* Interchangeable MI by Grado.

Just as Neumann and Scully lathes and Grampian and Cook cutterheads "scratch the surface" of a blank master lacquer, here we've only scratched the surface of many hundreds of electrical phonograph pickups and styli that evolved from the steel needles of a century ago. In light of vinyl's resurgence, it is a pitty that Stanton is a shadow of its former self. Also that Shure cartridges are no more. Vintage NOS (new old stock) is becoming scarce, and dear. New line contact styli are few, and even dearer. Ellipticals of 0.3x0.7mil (7.5x18μm) are the new normal, although they are inferior to line contacts in sound quality and wear. For the new generation of vinyl lovers, it is good that Ortofon in Denmark, Audio-Technica and Jico in Japan, Expert Stylus UK, and others hold to a rich tradition.

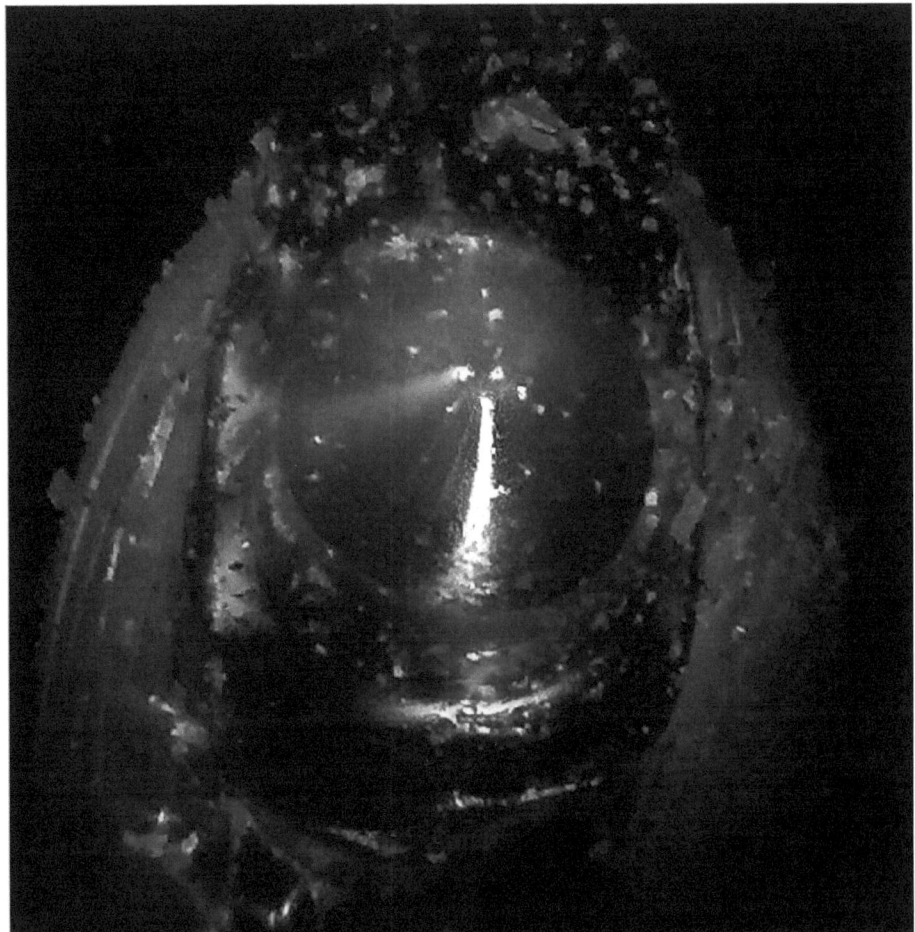

Decca-London 0.65 (16.5μm) nude spherical tip, between laminated magnetic poles in the previous image, surrounded by soft iron, tensioned by a loop of clear nylon "rope." With this complexity, the means of replacement is to return the pickup to an authorized retipper.

Apart from the stylus, the magnetic cartridge (pickup) body itself

In the 1950s of the earlier Pickering era, cartridge generators began to evolve from the MC (moving coil) era to moving iron (MI, aka variable reluctance or flux-valve). MI was developed by EMI, Fairchild, GE, Pickering, and in Germany by ELAC (Electroacustic GmbH). A slug of iron in the cantilever modulated the field of magnets in the cartridge body. After tiny powerful rare earth samarium cobalt (SmCo) magnets were possible, the moving magnet (MM) body had no magnet, only coil(s). Shure and Pickering in the US produced MM cartridges. Other methods were FM modulation (Zenith), variable capacitor (Weathers), and the four tiny coils on the cantilever of MC. $18k LASER players have no stylus so make no physical contact at all, but cannot play dirty or translucent colored disks!

With all magnetic pickups, one of three things must move. Either the coils must move, or the magnet, or a slug of iron "moves" to modulate a magnetic field. Most cartridges themselves, sans stylus, have no moving parts, so do not wear out. Only cantilevers move, and wear. Diamond styli can last 1,000hr, which averages to a thousand LPs, both sides! Used bodies can be had for peanuts. Audible preference between MM and MI is difficult to discern, or to defend. Were magnet weakening an issue, MM wins because a new magnet comes inside every cantilever. MM models are at both the low end, the mid-to-high range, and the very top of the product line. The mid and high range models have 4 coils for hum-bucking, introduced by Shure in 1962. But steel needles dominated for nearly 70yr!...

The first stylus – "soft," "medium," & "loud" steel needles

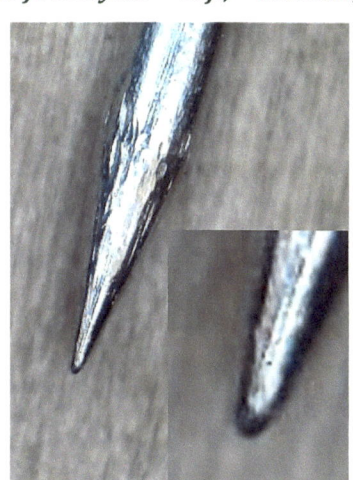 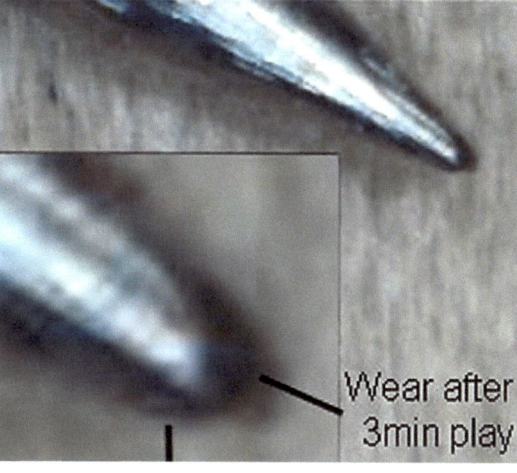

L: unworn Victrola steel needle. R: Wear after only one 3min side shows gray flats on both sides, ground by the very abrasive shellac disk. This stylus should not be used for the next disk!

Acoustic Victrola (**p105**) needles are steel ground hemi-spherical at the tip. Long shanks above the tip vary by diameter as "loud," "medium," & "soft," clamped in the linkage to the diaphragm of the "sound box" that drives a horn of metal above or folded in "shelves" of a wood box below, and with rotatable slats for volume.[15] The wide-groove disks were mono, cut vertically or laterally. Lateral cutting had the advantage of being "push-pull" by both groove walls; vertically the stylus pushed up by the groove, but free-falls by gravity, so even-order harmonic distortion is not cancelled. (WE and RCA tried a spring return, but it did not work well.) Disks were pressed in slate dust and the stuff of lac beetles, hence 78s are called "shellacs," though during WWII Bakelite was substituted, and eventually vinyl.

From the 1920s, electrically recorded disks needed to play compatibly on both the many existing acoustic reproducers and newly bought electrical players. Using primitive cutters with their own unique cross-section shape, standard play (SP) grooves were inconsistent disk-to-disk until well into modern electrical era of the last century. Macro-photos above show the 3mil steel stylus before and after playing a 3minute song at 78rpm. Within a few rotations of the lead-in, the highly abrasive 78rpm disks polished the needle to match the current groove. Smooth when new, now the edge around the stylus' side flats are sharp as the cutters! If the steel needle were reused for the next record, which likely had a different groove shape, it could inflict damage. Best advice was to use a fresh stylus for each side.

---

[15] Thorn and cactus needles were also tried with varying success, sold to restorers and as boutique items today.

Micro-groove for high fidelity monophonic & stereophonic sound

Vertical modulation continued for some broadcast ETs, and its coloration was touted by some, but lower distortion lateral cutting won for several decades of 78s, and a decade of mono LPs. Early stereo recorded the left channel laterally and the right vertically, but the quality difference was distracting. The solution in 1958 was to put Left & Right (L & R) channels on either wall of the stereo 45\45° **V**-grove invented by Blumlein at EMI in 1932.

The most correlated sounds (centered, tending monophonic) present as lateral motion, the *sum of L+R* signals. Uncorrelated sounds (side sounds, ambience, spatial reverberation) are *the difference of L–R signals* and are carried by inferior vertical motion. This is the reason lateral-only monophonic recordings from the same session sound less distorted than stereo, if less spatial. Now with the 45\45° groove, the higher vertical distortion is shared equally by *L & R* channels. Stereo plays with higher distortion, but as it is the same *L & R*, listeners perceive no difference between channels. No differenced takes away the reference for any inferiority. So there is no distraction. Of course, digital audio suffers none of this.

Phonograph styli faithful to the recorded groove

Through the revolution about 1925 of recording electrically and even into the 1950s, SP grooves were carved in wax, which was too soft to play. Musicians left the studio with no way to hear the performance. Unless rushed through plating and molding processes, artists & producers had to wait a week for a test pressing, maybe to return to the studio for retakes.

From the 1930s, immediately playable lacquered aluminum were *instantaneous acetates* and masters. Hydraulically pressed (stamped) vinyl plastic replaced shellac for distribution disks. Record lathes became sophisticated, their sapphire chisels tooled to tolerances of less than 1/10,000 of an inch (0.25μm). Long-lasting diamond play tips had to be ground and polished within these same tolerances. A big step toward faithfully tracing a groove.

Tip profiles on **p20** are illustrated below again for convenience. Styli may be either bonded, or a "nude" gem inserted in the cantilever, precisely aimed fore & aft. Whether polished at the factory for demanding hobbyists or after a "break-in" [16] of several dirty abrasive disks by the user, it is polishing that determines the stylus' final shape – and *how well it fits and traces* to reproduce the groove. Tracing the groove's modulation is at best imperfect. The replay errors produce artifacts. So next, we relate tip shapes to *distortion*.

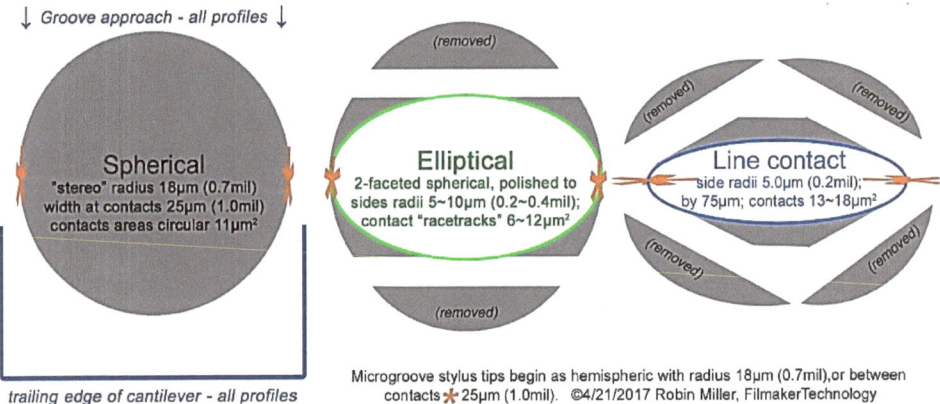

[16] *Breaking in* applies to initiating (loosening) mechanical cantilever elastomers & speaker suspensions, but not electronics beyond forming filter capacitors and thermally stabilizing electronics after 20min or so each time it's switched on (I leave solid-state on). Certainly not wires! Most "broken-in" is a listener's changeable perception.

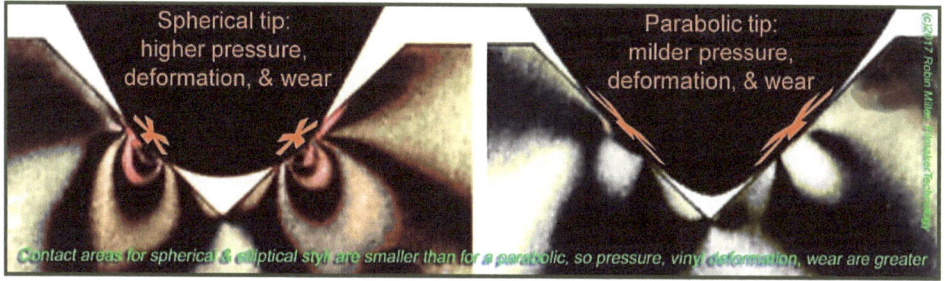

In these illustrations, orange asterisks (*) suggest the shapes and orientations of contact areas along and up\down the groove walls by stylus types. As discussed, medium area circles for spherical; concentrated ovals for elliptical. Large "racetrack" areas for line contacts (aka shield-shaped "parabolics") elongated several times their side dimension up\down both walls. Along groove motion, the tracing\scanning dimension is *shorter in time* to resolve high frequencies. Stylus footprints up\down the walls are made taller to spread tracking pressure over a larger area, reducing wear. Eventually tip wear damages the engraved groove. And accumulating groove damage sounds of ever more distortion.

Intense pressure of the passing stylus instantly softens and deforms the vinyl. Within seconds of the stylus' passing, the vinyl rebounds and resolidifies.[17] But the stylus' sides, tracing along groove motion, add distortion on every play from the first of a pristine disk, especially if skating is unequal on the two walls. Flats worn independently at the stylus contacts develop sharp edges around the flats that inflict permanent damage at the tracing contacts of the groove wall, akin to recutting the record. Unworn advanced profile styli in well-aligned arms can play the same disk dozens of times without discernible deterioration of the recording. But distortion is caused because *no stylus exactly fits the groove!*

Vectorscope images of the stylus' movements

How imperfectly the stylus traces the groove is what distorts (colors) the sound. Some hobbyists come to prefer (habituate) to this non-neutral reproduction. Brightness is super-sized; tone-color that is bigger than life-like is added on replay. Distortion artifacts are not recorded in the groove. Unless the producer intended a *novel* instrument.

Stylus motion left, right, up, and down as viewed from the front, is transduced to *L & R* electrical output signals of the cartridge, as revealed in the images. Rotated clockwise 90°, the *vectorscope* images **overleaf** show, rather than speaker direction, tip motion gyrating over a fraction of a second. For positive-going audio signals, the L channel groove wall moves right & up; the R channel wall right & down. To complete visualizing this crazy motion in 3D, the groove approaches from the back, while the stylus tries to follow.

In stereo, the two independent groove walls are only under the stylus. Lateral motion is push-pull by the walls, cancelling even-order harmonic distortion. In stereo the stylus is *only pushed upward* by groove walls, *falling back only by gravity* due to the vertical tracking force. Vertical motion is *single-ended* and subject to even-order harmonic distortion. L & R are comprised of both lateral & vertical motion.

---

[17] "Vinyl" (poly-vinyl chloride, PVC) is naturally clear, or died black or in colors, with 25~30% recycled PVC in the mix; premium labels use quieter virgin~10% recycled. Pressings vary 120~140g, "audiophile" grades 180~200g, with little effect on sound. Pressing stresses are relieved long after purchase, leading thick disks to self-warp!

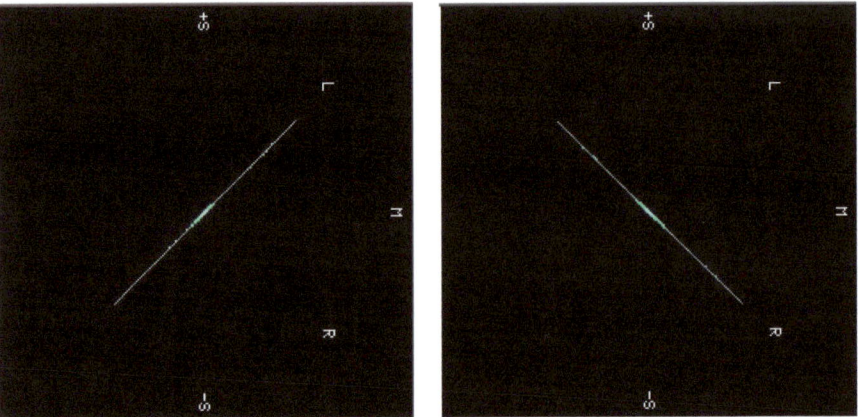

90°-rotated vectorscope images to show stylus motion, sequentially lateral (horizontal) and vertical. **Left,** viewing from the front, signal only on the **L** channel & groove wall (positive-going right- & upward). **Right,** signal only on the **R** channel and groove wall (positive-going right- & down-ward).

The sounds on only one groove wall are heard either from the L speaker, or from the R speaker, as above. Phantom sounds between speakers tend toward the *highly correlated* (near-identical events) of L & R, above called "M," for middle, that below-left tends to modulate the groove laterally. Stereo spatiality (side sounds, ambience\reverberation) is *highly uncorrelated*, random phase between L & R signals we call "S," for sides. Below "S" signals tend to wiggle the groove vertically. In any interval we find the stylus tip on a path, wildly changing direction. We describe it either form: "L-R" or "M-S" (mid-sides).[18]

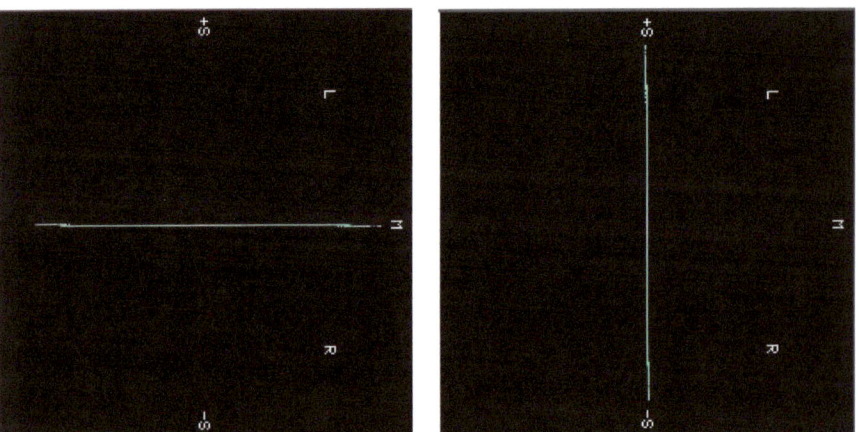

Left, lateral-only motion (mono, "**M**", positive-going is rightward). **Right,** vertical-only (mono out-of-phase, "**S**", positive going is upward). The groove moving any way rightward results from positive-going signals in-phase between **L** & **R** channels. The groove moving any way up & down, signals are out of phase.

Unlike monophonic reproduction with one speaker, stereo "pans" frontal auditory events at or anywhere between two speakers, represented on the vectorscope by "L," through "M," to "R." Recorded spatial ambience are "S" sounds heard off center and even *outside* the speakers. The cartridge extracts these two channel signals from the L & R groove walls.

---

[18] Complete expressions in M-S form: $M = \frac{1}{2}(L + R + (L_{VLF} + R_{VLF}))$; $S = \frac{1}{2}(L - R - (L_{VLF} + R_{VLF}))$. Cutter head drives (thus final groove wall signals) are $L = \frac{1}{2}(M + S) = L + (L_{VLF} + R_{VLF})$; $R = \frac{1}{2}(M - S) = R - (L_{VLF} + R_{VLF})$.

The *vectoscopes* below left shows typical 3D *stereo stylus* motion corresponding to sounds in time that are nicely balanced between correlated M and uncorrelated S of L & R signals in M & S form. Its slightly football shape results from the somewhat lower S signal from reducing LF differences between L & R to prevent the stylus hopping grooves. If not by panning center the electric bass and kick, it's by summing sounds below ~150Hz, SOP mastering in stereo, in effect redistributing LF energy from vertical S to lateral M.

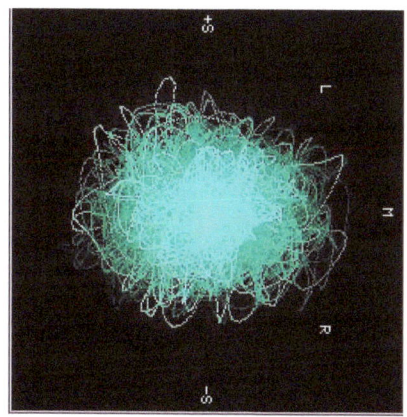 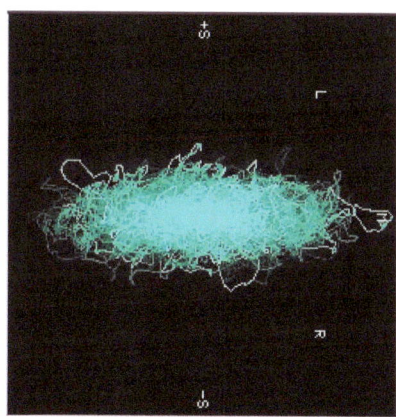

L, stereo groove motion any way rightward are positive-going signals, channels tending in-phase (as the sum **M**). Vertical motion "**S**" tend to be difference, random-phase. **R,** a monophonic record *using a stereo pickup.* What should be a horizontal line, any vertical motion, intentional in stereo, is distortion in mono.

The right image above reveals *replay of a monophonic record with a stereo pickup.* A mono signal *as cut* is only lateral (M), as in the vector display of a horizontal line, **opposite** bottom left. *For Mono play, all vertical motion S is distortion!* Despite using a fine D81Sii on a well-aligned SME-3009 arm. Easily cancelled by mono mixing in the preamp, whereby the display collapses to M's original horizontal line, the sound again is distortion free. But playing stereo disks, such added distortion artifacts remain, and are *heard!*

Distortions playing vinyl are stylus motion that is *not in the groove.* Illustrated **overleaf** as though viewing from under the groove, 2nd harmonics are generated mechanically on playback. Where the tip rides the groove walls of a simple mono sine wave, its **red** contact paths are *parallel radially* (not *tangentially*) with respect to the disk by the flat-across "Recording chisel," shown at peak and zero modulations. But tangential to groove motion, horizontally from right to left, the groove appears to narrow, widen, then narrow again, etc.

The "Spherical stylus" with circular cross-section fits and drops down where it can, at waveform peaks. But at the zero-crossings, it is pinched up to its highest. The narrow-sided "Elliptical stylus" or line contact tip fits better everywhere, so rises and falls less. A vertical "S" artifact has been added on replay. Again: *Not in the recording, it is distortion.*

Called "pinch effect" distortion, this rising & falling happens twice per recorded cycle as illustrated, adding a strong artifact an octave above the recorded tones. Mixing into the original sound *even-order harmonics* alters its timbre, "discoloring" the sound (making it overly "bright"). Worse, adding 2nd harmonic to many instruments that normally produce *only odd-order harmonics* (trumpets, flutes), their timbre tends to sound like violins!

At its worst, at the inner groove at highest levels, the pinch-effect turns high frequency sounds into raspy spit. Most noticeable on vocal sibilant sounds (S's, T's, Ch's,), it has led audiophiles to call this *curvature overload* distortion on replay "sibilance." *Opposite in polarity* between stereo channels, the *raspy spits fly around the room outside the speakers!*

Vectorscope images of the stylus' movements

Stylus unfaithful to the groove – mechanical distortion mechanisms

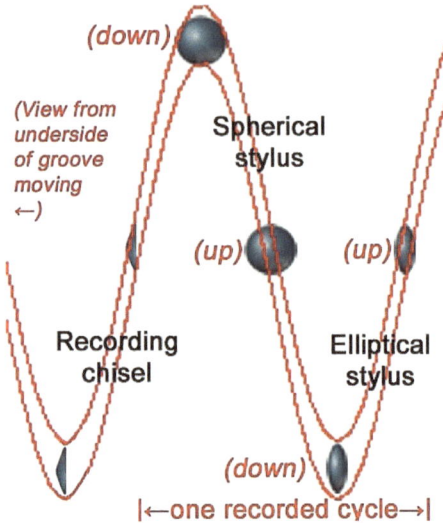

Styli create distortion that is *not* in the recorded groove. The groove is cut by a flat chisel, shown at a peak and mid-point of modulation. Spherical styli drop down at peaks, but pinch upward in the middle, adding a 2nd harmonic, opposite in polarity in stereo. Elliptical styli are not "pinched" as much; line-contacts least.

Considering this important pinch effect further, the illustration above is a lateral-only Mono groove (or the *correlated* center-stage sounds in stereo). The cutter has swung side-to-side without steering, so the walls are *radially parallel,* duplicate sides shifted radially. Not like the sides of a tiny road that would swing with the curves. But when played, like riding in a tiny round car, a spherical tip encounters the groove\lane narrowing as the tip crosses the middle. This constricting *pinches* the spherical tip up, falling back down at the next peak excursion of the modulated groove. To mono\correlated audio mastered only laterally, the stylus has added a vertical artifact that, with a stereo pickup, is pure distortion.

Narrower elliptical and line-contact tips rise and fall less, producing less pinch distortion. Proper play in mono mixes L & R to cancel vertical motion, silencing pinch distortion. But other distortion artifacts *not* silenced by the mono switch are termed *poid-like*. Illustrated **opposite**, p*oid* describes the path of a curved edge riding a sine wave surface: it scans a high frequency faster on the leading side than on the lagging side. Sharpening outside peaks of the wave and broadening inside peaks. The fatter the scanning edge (spherical, fat elliptical), the greater the poid effect; narrow ellipticals and line contacts create the least. *Poid* affects most high-level HF in the inner groove, transforming an original sine-wave (its groove walls' contact lines in **orange**) into a ***sawtooth*** waveform (**yellow** in inset).

Poid worsens when the tip is skewed, shown at a typical observed QC limit 7°. The R output is stretched from time **t1,** when on its trailing side the stylus is released more slowly. Then from **t2,** the tip slams into the oncoming rise and quickens in response, depicted in the yellow waveform of the inset. Instead of the recorded sine wave, the cartridges outputs a sawtooth, adding even- and odd-order harmonics.[19] While pinch effect adds more benign even-order harmonics, poid effect's sawtooth adds less musical odd-order harmonics, plus more even-order ones, adding to total harmonic distortion (THD) up to double digits.

---

[19] "Poid" distortion was first described in 1938 by Harvard researchers Pierce & Hunt, who in that publication also foresaw the benefits of using microgroove elliptical and line-contact styli. The 1930s were great for audio!

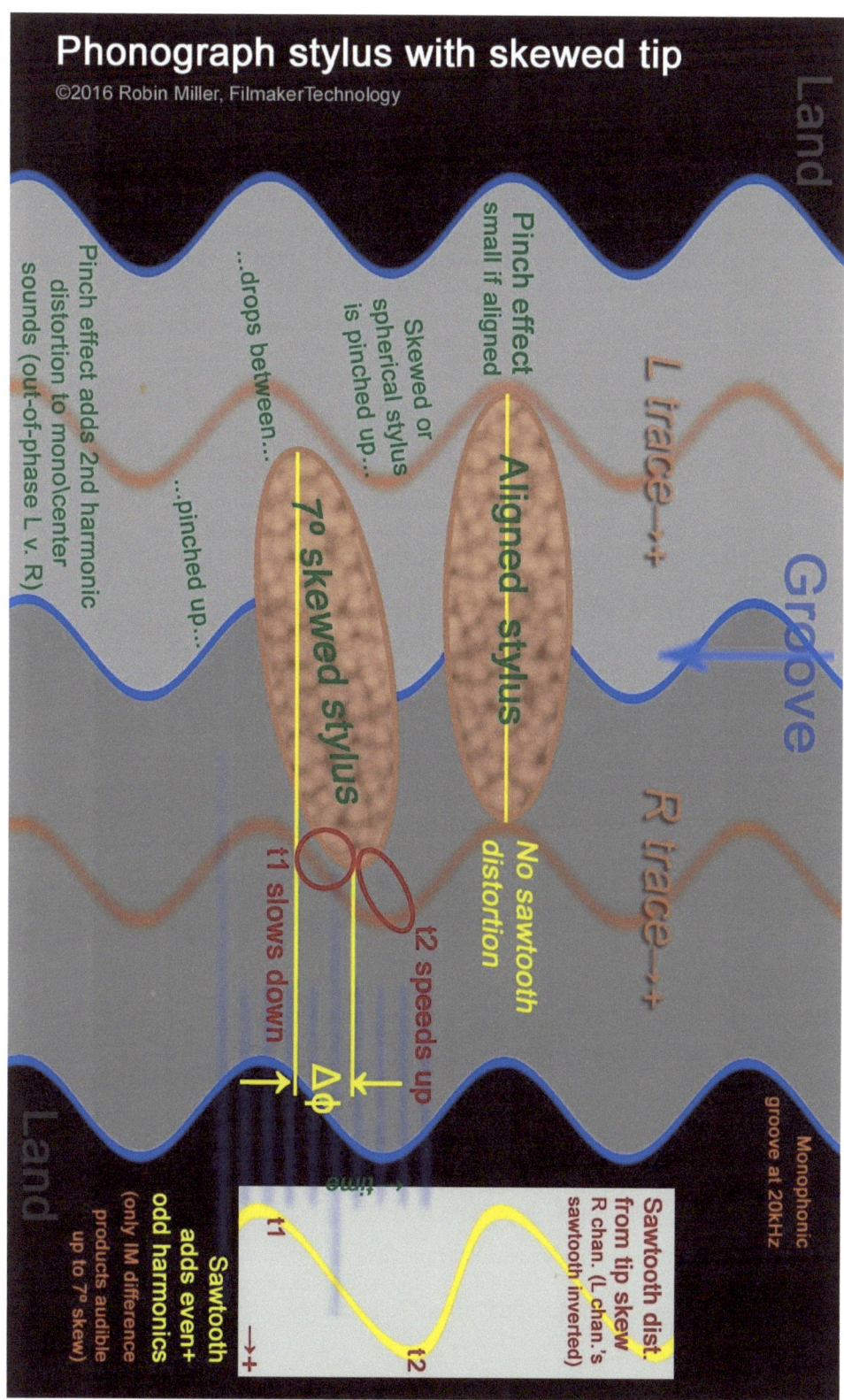

Stylus unfaithful to the groove – mechanical distortion mechanisms

Skewed off-course ≤7° or not, lower poid-like harmonics may be ultrasonic, therefore inaudible. But these non-linearities also produce comparable *intermodulation distortion*, or IMD, the lower sidebands of which can reach throughout the audible range, heard as non-musical "burr" sounds, especially at peaks of recorded sounds, in which they don't exist.[20]

Poid-like THD falls to single digits if the tip is aligned. A well-aimed tip sends identical distortion signals to the speakers for center sounds. But pinch and poid-like artifacts are opposite in polarity between channels, flashing instantly *outside* the speakers, so they smear off-center artifacts of the loudest, most important solo voices, causing them to lose "focus." Properly aimed, the tip's width and truncation prevent its bottoming. But if skewed, the tip descends farther down, reproducing clicks & pops of fallen debris ground in there.

Tip aim is a common manufacturing error in my experience, and problematic if more than ~7°. I rarely see it in older Plainview NY Stantons or Pickerings through the 1980s. But the company showed worsening craftsmanship and quality during its "transition." Recent styli I've acquired (and returned) have had more than 7° of tip skew, as I now appraise acquisitions, new as well as used, under a laboratory microscope. However a 60x inspection loupe can reveal this problem (available online with illuminator for ~$5).

Again, none of these distortion mechanisms apply to digital audio. While digital artifacts are usually milder, intentional clipping is worse than any vinyl distortion. The purpose of this book is than most distortion mechanism in vinyl can be softened by attentive users.

Replacement \ aftermarket \ generic styli – caveat emptor

For makers today, spherical styli are easy. Ellipticals are harder. Line contacts hardest. Sometimes trickery substitutes for the skilled labor required. To cut costs, some omit the tie-wire stabilizing the cantilever. Others are sloppy about tip skew, or bonding that fails. As should happen, reputable sources need no discussion, but buying online: buyer beware. Scam artists abound, some unknowingly mis-represent particulars; others commit fraud.

One dealer from whom the author purchased a line-contact stylus mailed a spherical not even for the same cartridge model line. Another charged market price for a D81ii featured earlier and claiming it "NOS in good condition:" the "sealed" item, shown here, had no tip.

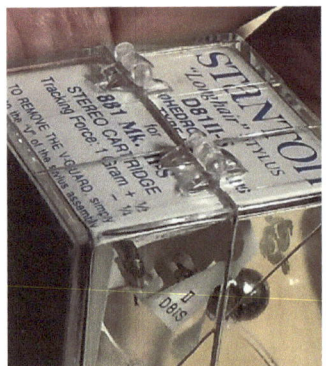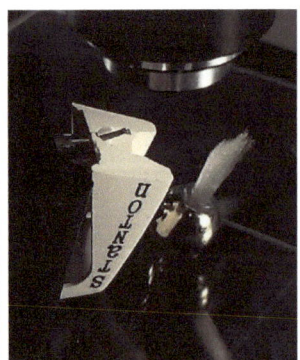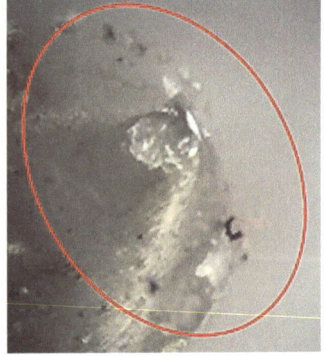

A find on ebay described as NOS…but under the microscope…it had a decapitated tip! *Buyer beware!*

Opposite, a Stanton reseller described an elliptical as "brand-new, unopened, undamaged item in its original packing." Not a lie *per se*, but its tip was rebonded with a spherical tip! **Overleaf**, "for" in an auction item title connotes *not original* "D6800eeeS" Stereohedron.

---

[20] All sounds are a time-varying harmonic structure that is the instantaneous sum of sine waves. [Fourier 1822].

An ebay purchase advertised as NOS, ostensibly a good choice of 0.3x0.8mil (7.5x20μm), useful for mono records. But under the microscope, this tip is NOT elliptical! The seller honored a full refund.

At the price of a genuine NOS line-contact, the tip **overleaf** is of unknown manufacture, packaged by otherwise reputable seller. To cut costs, this ("qd"=Quad?) Stereohedron replacement has two facets, not four. Laying on the surface of the cone in each plane might fit the "line-contact" definition *per se*. From both the reader's and the groove wall's point of view, the curved edge scans the groove obliquely *off* in vertical tracking \ stylus rake angle (VTA\SRA), not fitting the HF undulations. As on **p54~56**, it creates sawtooth THD by the groove attacking the tip's edge slowly from the bottom right, then releasing fast.

This tip sounds of artificial "brightness," "thin" in bass, spoiling tone color. It has a tie wire, and tracks well at 1.25g. But its 8.3° tip skew is troubling, adding further poid-effect, and allowing the tip to sink farther into bottom-dwelling debris, accenting dirt noise. By test disk or by ear, HF response could be equalized, but the distortion cannot be undone.[21] Not rounded by polishing, the sharp edge damages grooves right from new. "Generic" styli need this microscopic sleuthing to identify deleterious cost cutting that only fattens profits.

---

[21] Later, RIAA "equalization" by the user's preamp restores flat frequency and phase responses to the original.

Replacement \ aftermarket \ generic styli – caveat emptor

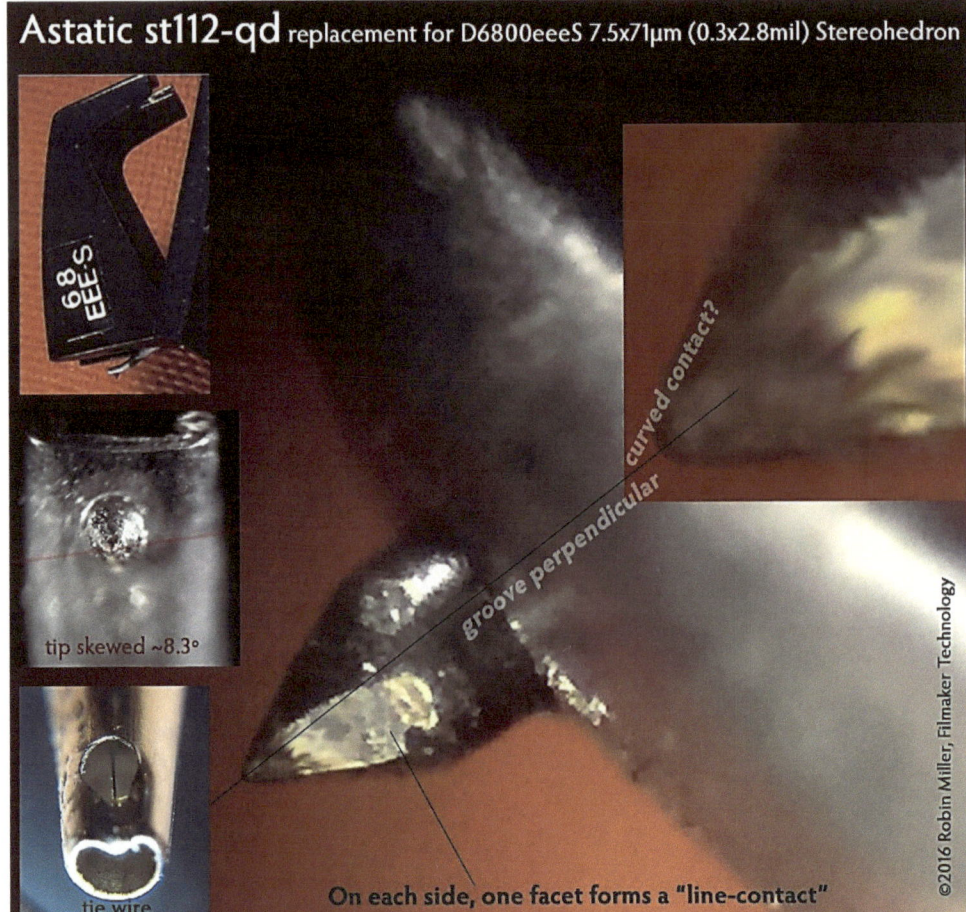

Astatic st112-qd replacement for D6800eeeS 7.5x71μm (0.3x2.8mil) Stereohedron

S-EEE 68#

tip skewed ~8.3°

tie wire

groove perpendicular

curved contact?

©2016 Robin Miller, Filmaker Technology

On each side, one facet forms a "line-contact"

How the stylus & partners re-construct sound

This book is *not* a shopping guide to the myriad makes and models of "turntables." It *is* knowledge for readers to make good choices. As explored, the "sound of the stylus" is its coloration (distortion) tracing the sound in the groove. Permanent "distortion" is from groove damage. External to the cartridge, the tonearm must be seen to re stylus alignment, tracking pressure, skating compensation, and LF resonance. The "spinner" determines the rotating speed (pitch, wow & flutter), and rumble vibration. And the preamp sets cartridge loading (HF response), accurate EQ, reduction of rumble, further reduction of vertical rumble in stereo, stereo "soundstage," distortion cancellation in mono and LF in stereo.

Integrated audio recording systems, such as magnetic tape machines or digital sampling devices have mechanics and electronics designed, manufactured, and tested as a unit. But phonograph systems may be individual components cobbled together by the user\installer. Even if integrated, a turntable's chain of individual links can interact in ways that cause further trouble. Our story continues in the broader context of the phonograph *system*.

Ideally the choice of cartridge and stylus has considered the types of records and music genre to be played, turntable functionality (manual to automatic changer), the intervening tonearm, the following electronic preamplifier ("phono stage"), and the over-riding budget. Even for modest systems, interacting dependencies can be optimized for best sound quality.

Most intimately mated with the cartridge+stylus are the tonearm and the preamp. Usually these are set up once as a system by a professional installer or studious consumer. If the pickup is not properly integrated with its arm and preamp, none will be worth their costs!

Alignment – a pickup cartridge mechanically mated with a tonearm

The stylus is the mechanical component of the phono pickup (cartridge), which in turn is the electro-magnetic contributor. Some tonearms move by a *linear*-tracking (tangential) mechanism that emulates a record cutting lathe. More commonly, arms track the record's spiral groove pivoting from a fixed point. These use a precise geometry, below left, to minimize *tracking error distortion* that compounds the stylus tracing issues. Tangency error less than one degree is possible, especially with longer arms, by optimizing two key measurements: 1) cantilever' *offset angle* set using an alignment "protractor;" and 2) tip *overhang,* checked by the gauge below right, or the protractor, such as the one **overleaf**. DJs scratching vinyl use *no* offset angle or overhang, which adds considerable distortion.

Highest sound quality calls for users\installers to align properly the cartridge in its tonearm. A full alignment with a protractor takes about 20min and is needed for any change past the tone arm pivot. Periodic alignment checks take 20s. Misalignment causes tracking distortion (%THD) and a comparable measure of intermodulation distortion (%IMD), as graphed on **p61** for a common 9in (225mm) arm once optimally aligned.

Overhang is how far the stylus tip extends past the spindle. It is the effective tonearm length *pivot-to-tip* minus the *pivot-to-spindle* distance. The **offset angle**, ranging from 15° for a broadcast 15in (365mm) arm to 28° for an 8in (200mm) arm, is by a bend in the arm, or by skewing the cartridge in an otherwise straight arm. Below right, a cutout for checking overhang that can be glued to a 45 adapter, downloadable in this book's Updates at www.filmaker.com/papers/UPDATES_RMiller-Better Sound of the Phonograph.pdf. [22]

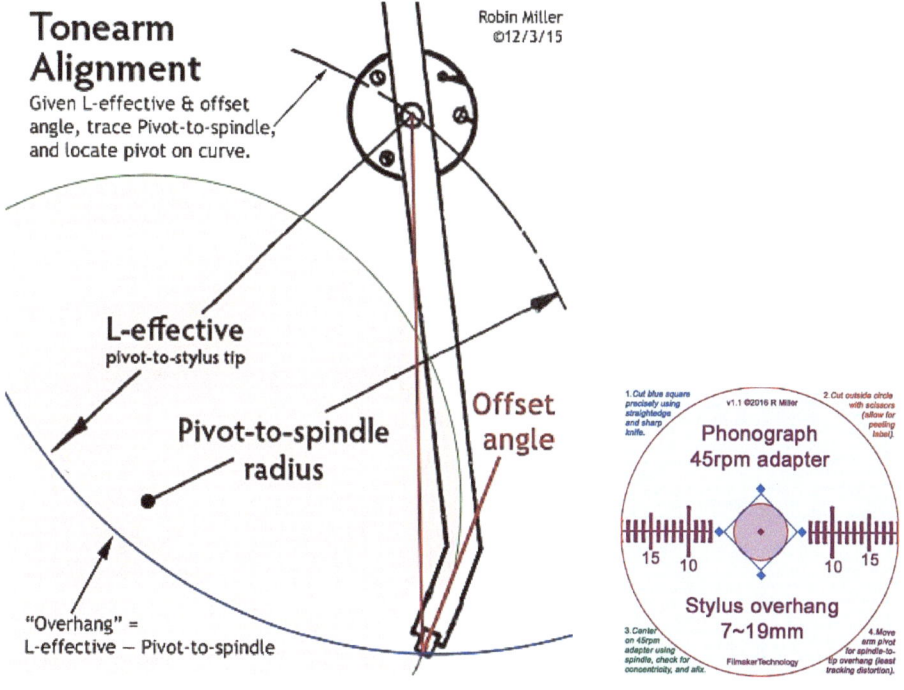

[22] If the arm cannot swing to the spindle, overhang simply equals *pivot-to-tip* minus *pivot-to-spindle* distances.

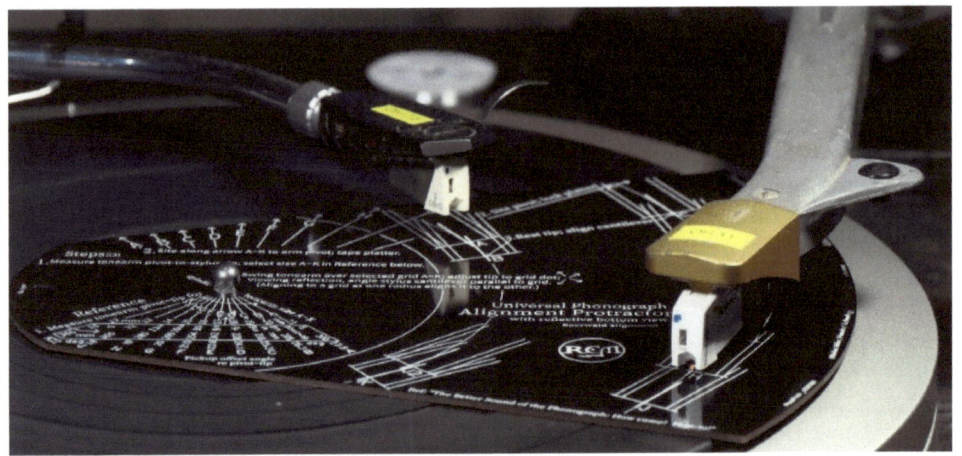

The author's Universal Phonograph Alignment Protractor (UPAP) with mirrored surface to view the cantilever from underneath the cartridge for aiming precisely.

A *protractor* is used to set or quickly check tonearm alignment. After choosing the alignment protocol and disk diameter, almost any protractor's instructions in general are:

1. Gently place the stylus tip near one of the two grid targets;
2. Sighting from the front, confirm no dogleg between cartridge and its reflected image;
   a. To adjust, twist a removable headshell, or shim under the cartridge mounting ear;
3. Slide cartridge in the headshell until the tip reaches the target dot to set *overhang;*
4. Viewing from directly front using a magnifier or phone camera, bounce a light off mirrored protractor surface to the underside of the cartridge, focusing on cantilever;
5. Skew (yaw) cartridge to aim the cantilever's *offset angle* over the target's centerline;
   a. Confirm that the centerline obscures the cantilever equally side-to-side;
6. Recheck the overhang (tip on dot).

Tonearm alignment table [mm,°]						Löfgren-A (Baerwald)					©2017 Robin Miller, FilmakerTechnology
Effect. Length	Offset Angle°	Over-hang	Pivot to Spindle	maxTHD 12in LP	Skatg F % VTF	Effect. Length	Offset Angle°	Over-hang	Pivot to Spindle	maxTHD 12in LP	Skatg F % VTF
200	27.8	20.9	179.1	0.78%	16%	228	24.1	18.1	209.9	0.66%	13%
202	27.5	20.7	181.3	0.77%	16%	230	23.9	17.9	212.1	0.66%	13%
204	27.2	20.4	183.6	0.76%	15%	232	23.7	17.8	214.2	0.65%	13%
206	26.9	20.2	185.8	0.75%	15%	234	23.5	17.6	216.4	0.64%	13%
208	26.6	20.0	188.0	0.74%	15%	236	23.3	17.4	218.6	0.64%	13%
210	26.3	19.8	190.2	0.73%	15%	238	23.0	17.3	220.7	0.63%	13%
212	26.1	19.6	192.4	0.72%	15%	240	22.8	17.1	222.9	0.62%	13%
214	25.8	19.4	194.6	0.72%	15%	242	22.6	17.0	225.0	0.62%	13%
216	25.6	19.2	196.8	0.71%	14%	244	22.4	16.8	227.2	0.61%	12%
218	25.3	19.0	199.0	0.70%	14%	246	22.3	16.7	229.3	0.60%	12%
220	25.1	18.8	201.2	0.69%	14%	248	22.1	16.5	231.5	0.60%	12%
222	24.8	18.6	203.4	0.68%	14%	250	21.9	16.4	233.6	0.59%	12%
224	24.6	18.4	205.6	0.68%	14%	252	21.7	16.2	235.8	0.59%	12%
226	24.3	18.3	207.7	0.67%	14%	254	21.5	16.1	237.9	0.58%	12%
Compared to steampunk 12in (305mm) tonearm project:						305	18.0	13.0	292.0	0.47%	10%

Alignment charted for tonearms with Effective Length (pivot-to-tip) of 8~10in (200~254mm). Compare a consumer 225mm (9in) with a 305mm (12in) transcription tonearm, for which maker instructions come later.

Consumer tonearms ranging 200~254mm (8~10in) effective length, pivot-to-stylus tip, are tabulated **opposite**, in contrast to a transcription-length 305mm (12in) tonearm (0.47% maximum), such as the maker project later in this book using ordinary hardware. Each size has its own optimal offset and overhang for proper alignment, and maximum tracking THD.

The specific tracking THD v. dimensions are **below** for typical tonearm 9in (225mm). With this dimension and for a 12in LP, a proper alignment using the *Löfgren-A (Baerwald)* protocol produces a compromise alignment trajectory that yields total harmonic distortion (THD) due to mis-tracking of 0.67% at three points: the outermost groove radius at 146mm, a groove radius of 84mm, and the innermost groove radius of 60mm. Precise tangency with zero distortion occurs at null points of 65.7 and 121.7mm. [The author provides clients this customized spreadsheet for their choice of tonearm and disk format.]

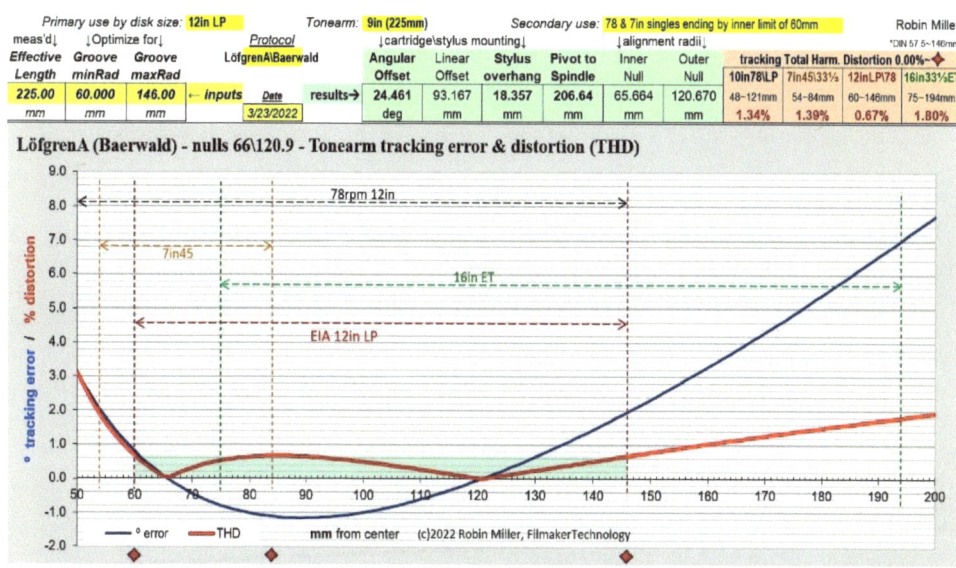

A 9in (225mm) arm produces 0.67% maximum THD at three points, to 0% at two points across the disk. Maximum mistracking distortions for four record formats are in **red type,** ranging for 12in LPs in the red curve in green region. Data entered in **yellow boxes** are calculated alignment distances, **boxed in green**.

Other alignment protocols, disk sizes, THD results, and protractors are in this book's **Update,** linked within the Phonograph book description at www.filmaker.com/papers.htm, or download at www.filmaker.com/papers/UPDATES_RMiller-Better Sound of the Phonograph.pdf .

Vertical tracking force (VTF)

Before your finished aligning, confirm correct tracking force (VTF by the tonearm's counterweight), skating compensation, and resonance in the next sections. In addition to supporting the cartridge at its proper alignment and permitting it to travel across the disk unimpeded, the tonearm provides VTF with the range specified by the maker for the stylus, historically in ounces (1oz=28.3g), but today typically 1~3g. Expert Stylus UK advises a maximum of 3g to avoid permanent vinyl groove damage. Medium- and high cantilever compliance dictate a lower VTF of 1~2g, and an arm of lower "mass"[23] Later, *compliance & mass* also *resonate* causing flabby (ringing) bass, speaker feedback, and groove jumping.

---

[23] Or *angular momentum,* which is *moment of inertia* × *angular velocity* as the groove pries the cantilever mass.

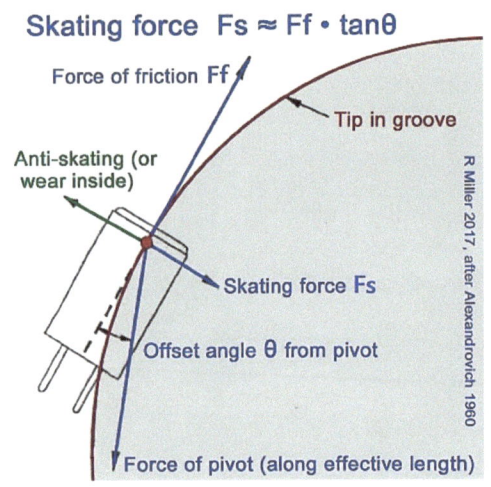

L, skating due to *friction Ff* & arm *offset angle*. **R**, a weight to set unvarying anti-skating torque.

Causes of "skating," and adjusting anti-skating (bias)

Friction between the tip and the vinyl groove tugs at the tonearm, which because of its offset angle, pulls the tip toward the spindle, called *skating*. The tip is pressed on the inner groove wall, thus pulled away from the outer wall. The force vectors are shown above. Without compensation, uneven wear of the stylus and groove result, and distortion on the right channel by the tip intermittently losing contact with the outside wall. Skating force redirects ~13% of VTF, now lost from pressing the stylus downward in the groove, causing reduced tracking ability and groove hopping. *Uneven tip wear accelerates need for stylus replacement by limiting tip life to the more worn side, e.g., reducing life from 800 to 400hr.*

Anti-skating ("bias") is needed, but if set too high, distortion becomes greater on the left channel.[24] Sliding friction and therefore skating vary very little with HF energy in the groove, and not at all with the linear speed of the groove from beginning to end of a side. Vinyl coefficient of friction (~0.30)[25] is the dominant variable, so we needn't consider lesser factors further. But as vinyl friction varies disk to disk, skating force and therefore the need for anti-skating is variable. Although skating force is unpredictable, anti-skating is usually applied unvaryingly by a spring, or by a thread, pulley, and weight, shown **above**.

Better than setting the tonearm's one-size-fits-all anti-skating is to give each disk side two quick observations, one *visual*, one *audible*. 1) If while playing the disk the cantilever is pulled off-center toward the outside, the culprit is skating tugging the arm toward the spindle; if the stylus is off-center toward the spindle, then anti-skating is set too high.

---

[24] Higher with greater offset angle, 12in arm skating is ~73% of a 9in arm, part of the resulting force of vinyl sliding friction (Ff above), equal to the tangent of the offset angle (0.445 for a 9in arm) in the spindle direction. Lower effect means longer arms also reduce skating force varying with differing vinyl compositions.

[25] Sliding (kinetic) friction is the predominant and only constant cause of skating. Results in loss of track-ability, uneven stylus & groove wear. Other highly variable factors are insignificant: groove radius (±5% of the friction effect) and high-level HF content [Kogen, Shure 1967]. Coefficients of friction for a spherical tip tracing vinyl vary 0.15~0.43+ with vinyl composition used with the label [Pardee 1981], and compare to normalized "wear" in the spreadsheet on **p22** (more for elliptical, less for line-contacts). The anti-skating compensation called for is only a bit more than the tangent of the offset angle times the force of friction, which is the coefficient of friction times VTF [Alexandrovich 1960]. For an 8in arm with offset ~28° average skating force alone is 0.158*VTF (~16% of VTF); for a 12in transcription tonearm with offset ~18°, it is 0.097*VTF (~10% of VTF), or ~62% of the 8in arm.

Audibly we are aware of distortion if for correlated (mono) sounds it differs between speaker channels; the channel with lower distortion serving as a reference for cleaner audio, so: 2) If peaks of loud centered content exhibit more distortion in the R channel than the L, then skating force is tugging the stylus away from the R (outer) groove wall; anti-skating is too little and should be increased. If distorted peaks are higher on the L, the stylus is being pulled away from the L (inner) groove wall, then anti-skating is too high. Bias is correct when no difference in high level HF distortion is heard. *But again, it changes disk to disk.*

For shellac's far harder surface, 78rpm skating is nearly a non-issue, as its coefficient of friction is ~1/5 that of vinyl 78s & 45s. Styrene 45rpm friction and skating is half vinyl's.

Cantilever-tonearm mechanical resonance

Another bit of alignment minimizes audible rumble, muddy bass, and mis-tracking, in the extreme causing the stylus to jump the groove. This is accentuated by infrasonic (<20Hz) *resonance,* caused by the stylus compliance (stiffness) interacting with the "mass" (moment of inertia) of the tonearm.[26] Imagine it as a weight (arm) pulled down to bounce on a spring (elastomer). Resonance should be tuned between 10~20Hz, between rumble and record warp <10Hz and recorded content >20Hz. Some say 8~12Hz. Integrated turntable-tonearm makers presumably addressed this issue, *but only for a single stylus' compliance.*

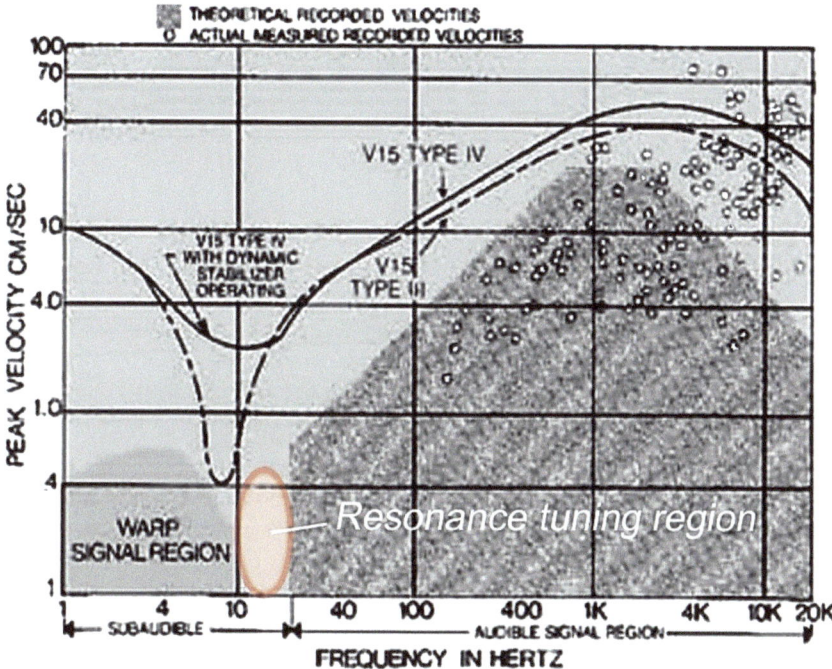

Courtesy Shure, resonance is tuned in the energy gap 10~20Hz to avoid amplifying disk warp & turntable rumble below 10Hz, and avoid boosting content >20Hz. Dark shading tops ~20cm/s crests, circles ~32cm/s reveal content dynamics 12dB above 8cm/s modulation crest, =5cm/s RMS RIAA reference for lateral mono, or stereo reference 3.54cm/s RMS. Circles show highest crests above 70cm/s, requiring 20+dB preamp headroom to avoid clipping.

---

[26] Analogous to a tuned circuit resonating at a frequency $f_r = 1000/(2\pi \cdot \sqrt{eM \cdot CU})$, where eM is the "effective mass" (moment of inertia) of the arm in grams acting as the tuned circuit's inductance, and CU in um/nN acts as the circuit's capacitance. E.g., a new D81 with CU of 30 on an SME-3009ii with eM of 9.5g, f_r=9.4Hz (increasing with elastomer aging). To lower the resonance frequency for a given stylus' compliance, increase arm mass.

How the stylus & partners re-construct sound

To support the arm + cartridge at VTF, a stylus' vertical compliance is less by the lower freedom allowed by the tie wire from the back of the cantilever to the back-top of the mounting tube (p15). The cantilever swings more easily laterally. If vertical compliance were half, then vertical resonance would be double the lateral frequency. *A second VLF bump an octave higher would likely be within audibility,* and furthermore *is out of polarity L v R!* But recall during disk mastering, vertical (difference) signals below ~150Hz are mixed to monaural (lateral), leaving no VLF vertical modulation (except disk warp <10Hz). In some cases, the arm can be given different moments of inertia horizontally & vertically. Archivist-grade preamps add low pass mixing at ~150Hz to null further vertical rumble.

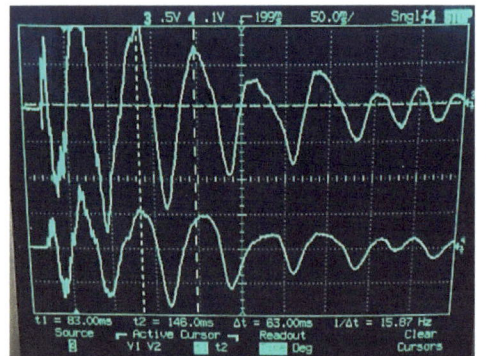 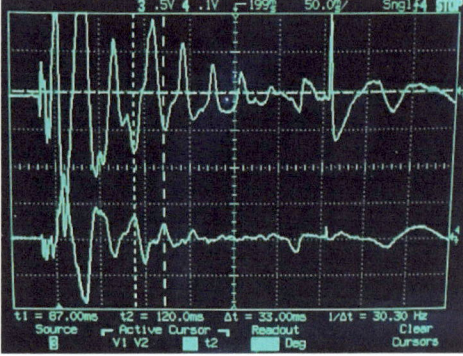

L: a rap (impulse) on the side of the turntable triggers a lateral resonance (impulse response, IR) in-phase at 15.87Hz, within the range that avoids rumble & muddy bass. *R:* rapping on top, a vertical resonance of 30.3Hz avoids disk warp, and the stylus jumping the groove, well out of the way of content.

Measuring resonance can be done with the stylus sitting still in a silent groove of a disk, and using the free digital app Audacity, if not an expensive Digital Sampling Oscilloscope (DSO). Either makes a short digital recording to capture the cantilever-tonearm system's response to a sharp rap on the turntable base with a knuckle or rubber handled screwdriver. The rap is called an *impulse*; the recorded "ringing" is called the *impulse response (IR)*. As in vectorscope images on **p52**, a rap on the side of the base produces a lateral in-phase IR. A vertical rap produces an IR out-of-phase between stereo channels. Either appears on the oscillographs **above** as nearly a sine wave that ideally diminishes within a few cycles.

Even new, resonance varies in manufacturing tolerances of styli with the same specified compliance units (CU), then its frequency rises with age. The elastomer's composition plus storage conditions might stiffen this rubber donut a little, or a lot. For an un-played new old stock (NOS) stylus, measuring resonance can be a forensic measure of its age. Or any stylus' CU can be accurately measured by comparing the stylus' resonance to the resonance of another stylus of known compliance. The IR, using Audacity or a DSO, measures how a stylus' CU might have changed. New, the D680 stylus in the oscillographs likely met its ~18CU compliance specification, and with a low mass tonearm resonated at ~10Hz. Over time, expect to add to head-shell "mass" to maintain resonance within the target range.

Vertical resonance may differ from lateral, typical by double the frequency. If still well below 150Hz, where vertical content has been filtered, there may be little consequence. *Monauralizing* replay below 150Hz again in the play preamp also cancels vertical rumble.

The stylus encounters a century and a half of dirty filthy disks

This book is one of the author's more comprehensive writings on audio – other papers are at www.filmaker.com/papers.htm. Later are maker instructions for an RIAA phono stage and a 12in transcription tonearm, both for low money but high satisfaction. Because

also popular again, along with vintage and new "vinyl" and appreciation for better sound, among the young as well as seasoned enthusiasts, is renewed enthusiasm for maker projects like the *kits* of the hi-fi era of the 1950s, when the author got bit by the bug. Playing phonograph disks can be fiddly at any level. Enthusiasts need greater working knowledge, and then to practice greater care than for far more convenient digital media. Not only attending to turntables, tonearms, cartridges, and styli, but caring for and cleaning disks.

Disks should be stored standing upright, not piled laying flat. Before playing, remove dust from the record surfaces before it becomes ground in the groove to cause clicks, pops, and accelerate groove and stylus wear. Easiest is a microfiber cloth dampened in filtered water and a drop of Dawn, then rinsed, all without getting the label wet. Most involved is a heated ultrasonic bath, shown **below**, using distilled water and dilute surfactant e.g. Dow industrial Tergitol. There is nothing to be done about an overly warped, if not melted, vinyl album left in a hot car or attic. Also irreparable, a scratch caused by careless handling of the tonearm. If not as a DJ back-cueing, a deep scratch then likely decapitated my son's tip **below**, leaving only its post with some bonding glue residue*!*

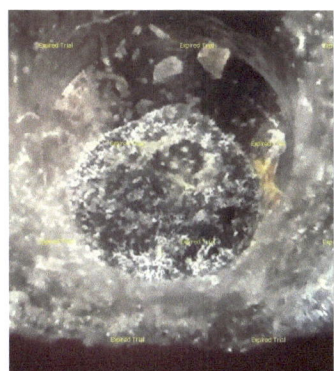 

L: A diamond heist? *R:* $5 60x inspection scope for tip skew and profile, but wear requires 200x++.
Below: An ultrasonic cleaning station: 2.6gal (10ℓ.) bath for up to 12in disks, and a jewelry-size for styli.

 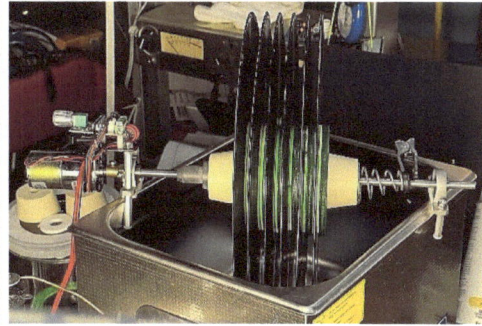

Played dirty, disks & styli accelerate their own demise. A disk damages a stylus, which subsequently damages disks, etc. The stylus grinds dirt more permanently into the groove, and the cycle escalates. Disks ought to be cleaned before each play. Clean only slightly dusty LPs with the velour covered "DiskWasher." For 78s, use a stiff-bristled brush and mildly soapy water, being careful not to douse the label. Filthier disks need immersion in ultrasonic water cavitation baths (**preceding image**) using dilute surfactant to dislodge the crud. The stylus should be inspected before playing. Worst case, it might need a few minutes in a jeweler's ultrasonic cleaner. Or routinely before each side, dab the tip gently a few times straight down in a bit of MagicEraser, kept handy atop the turntable.

How the stylus & partners re-construct sound

The "spinning" part of the turntable

The term *turntable* is often conflated to include a tonearm. More precisely it's the *spinning* component alone, as at the highest quality, arms are separately chosen & installed. Often the least problematic component, the spinner's maintenance is summed up in a drop of oil now & then, or a fresh rubber idler or belt. Spinners can be a source of mechanical distortions. Or if integrated with a tonearm, of any failure of its complex semi-automatic (returns to rest at the end of a side) or fully auto (that also cues to the start) mechanisms.

Today spinners are mostly one of three types: mechanical idler (puck) drive, belt drive, or electronic direct drive. Pucks develop flat spots, and belts stretch, so these rubber parts require periodic replacement, or their speed variations will *frequency modulate* (FM) the sound. Idlers replaced the long-ago gear drives of Vitaphone and broadcast transcription turntables for their dead-on timing using line-synchronous motors, at the price of vibration from meshing gears. Belts (sometimes strings) are audiophile favorites, also of the Library of Congress for archiving. The intermediary idler diameter means nothing: the speed is the ratio of the motor shaft capstan and driven platter rim diameters. Same with belt length.

Long as its electronics endure, direct drive (DD) is robust, high performing, and probably the most available used in good working condition. "Resolving" speed precisely to an unvarying drive frequency oscillator, DD turntables do not lose rpm to stylus friction, as can idler and belt spinners when rubber begins to slip and stretch. Recent models range between $179 to $4,000 with an integrated tonearm (still requiring a cartridge & stylus).

Mentioned in the section on resonance, the spinner must have low mechanical vibration (rumble) that might be picked up by the stylus through the platter as though it's in the recorded groove. A barely audible rumble spec for the turntable opposite is –73dB below nominal signal (–86dB for its best sibling is sensational, actually "unsensational" hearing-wise). For vertically recorded monophonic disks, or for stereo without LF mixing in the preamp, vertical vibration of the platter, idler, and bearings is as important as side-to-side.

The spinner must provide accurate rotation at two, three, or four standard speeds of 16⅔, 33⅓, 45, 78.26 in 60Hz countries or 78.92rpm in 50Hz countries. Off by an achievable and oft specified maximum deviation in speed of ±0.3% is out of tune by 5 "cents," 1/20 semi-tone, at the limit of what is perceivable in pitch (but ±4s off for a side). And must be low in speed variations: low frequency *wow* & high frequency *flutter*. An inaudible wow & flutter spec is 0.025%. For all drives, a massive platter acts as a flywheel to filter wow & flutter.

Digital audio can be marred by clock noise, digitization noise, and digital mixed with analog power supplies. With vinyl, electro-magnetic interference (EMI) from the motor can also be induced in pickup coils and tonearm wiring, hum from a ground loop, and radio frequency interference (RFI) from fluorescent lamps, dimmers, radio stations, or anything digital. Speakers too close to the disk, acting as a microphone diaphragm, can cause acoustic feedback, worsened by tonearm resonance already discussed.

Varying standard speeds is sometimes needed, especially for so-called 78s (65~90+rpm prior to standardizing. Easy for direct drive spinners with variable oscillators, switching to precisely "quartz-driven" at fixed speeds. The platter carries multiple magnets of a motor rotor around electronically driven stator poles. Some have an rpm display, and like many spinners have a stroboscope embossed on the platter rim illuminated by a pulsing lamp.

Direct drive turntable models for DJ use attain speed and wow & flutter spec in <¼ rotation, and some have brakes to stop on a dime. New model spinners cost $150~4,000 except for *über*-pricey works of art. Models with life remaining are readily available used in yard sales or at auction. Evolved as they are, we need say little more about spinners other than their two key performance artifacts, rumble, and flutter.

Below is a professional quality no frills DD spinner-only, often at auction for several hundred, still working like new at ~40yr old. The only maintenance is a quarterly drop of oil. It's simply mounted on two glued layers of 3/4in (19mm) plywood, on a solid cabinet (on a concrete floor). With tonearms for multiple disk formats. The preamp selects which arm\cartridge, and between critical *capacitive loading* (C-load, explored later) for each pickup. It's predecessor 3-speed shown on **p81** has even higher performance, and is still a workhorse, from radio to today's archival vinyl ingestion to digital. Lowest-end $99 all-in-one record players, with preamp and USB digital converter, have little of the quality sought in this book, so are not even attractive for their price.

Fine new & used spinners are siblings of this broadcast DD Technics SP-25, including the SL1200_ series popular with party DJs and audiophiles. Playing like new since 1982, this one has at back an SME 3009 tracking at 1¼g; at right, a 15in Audax broadcast arm tracking at 2¾g, both with line-contact styli.

Minimal accessories augment a turntable: a dense rubber mat; a massive base ("plinth"); a bubble level & means of leveling the base; a stylus pressure gauge. Only for warped disks is a record clamp that is much lighter and easier on the platter bearing than a record weight. For playing select cuts, soft lighting is the best, not a point source that creates specular reflections.

"Why bother, I can't hear it!"

This audio engineer hears this justification more than others by non-enthusiasts who claim not to care about audio quality. It's not that they dislike music. Or have never enjoyed a live concert. Given this book's cost sensitivity, it is unlikely a matter of $$. What is it?

Besides getting a hearing test, you might conduct a simple experiment to rate your *audio acuity*. For this experiment, you need time driving listening to a car radio, tuned to a distant stereo station, preferably playing relatively unprocessed acoustic music. While driving, do you hear the radio automatically switching between stereo and mono for best reception? In mono, the sound noticeably changes *character:* Noise & distortion decrease, but the *spatiality that is stereo* collapses. Perhaps hearing this, you *can* appreciate sound that is cleaner of artifacts, but alternatively enjoy feeling "transported" by the more life-like two-eared stereo. Aspects of "phonography" that relate and are explored in this book include stylus shape (profile), alignment, pinch effect, poid-like saw-tooth, tonearm, resonance, cartridge C-load, channel balance, accurate RIAA, clean disks, well-made recordings…

Striving scientifically for quality that eliminates distortions of several kinds, audio design engineers accept that, using test & measurement equipment, a device that measures *good* might still sound *bad*. But it is also true that a device that measures *bad* will almost never sound *good*. Yet some (not all) "audiophiles" prize hearing *coloration* (distortion), and compare components by trial & error for sound they like better than their last, or the other guy's. Never mind that audiophiles commonly disagree. Is what they prefer distortion? Which form? The result of random cancelations of systems component errors, as Norman Pickering cautioned? Have they forgotten an instrument's live *reference* sound? Become conditioned to a false sound of it? Perhaps they prefer a novel sound not found in nature?

Moving on from the turntable to the preamp…

We are about to end this book's exploration of what in popular parlance is called a "turntable," inclusive of an integrated spinner, tonearm, cartridge, and stylus. At highest quality, these are sought and bought separately, but aligned together for optimum replay quality. The phonograph *system* is completed by an RIAA preamplifier, up next.

Of these components, we've dived most deeply into the most critical part to get right, the stylus. Because of its several distortion mechanisms tracing a mechanical groove. In one of three shapes, it is the key part in audio performance, and in wear of both itself and the disks it plays, as tabulated in the spreadsheet on **p22**, and summarized more briefly on p34.

Cartridges have no moving parts other than the stylus tip and cantilever, and last pretty much forever if a coil inside doesn't happen to "open" (a fine coil wire break). Tonearms need alignment with the stylus' cantilever. And setting of tracking force and anti-skating. The spinner must be maintained for minimal rumble vibration and constant rotating speed.

Digital devices need none of these, and caution "no user-serviceable parts inside." The phonograph invites users inside for myriad optimizations upon installation, and periodic checks thereafter. And more will be needed to interface with the phono preamp, up next.

A reminder: this is all about the music, and spoken word, and natural sounds captured in 140+yr of historic recordings, archived in phonograph disks, and mostly not available in digital form. And is intended for readers ranging from archivists, possibly to preserve disks digitally, to diligent phonograph enthusiasts, whether new to the hobby or at it a while, including those for whom this book will be largely a review of what they already know.

Now with the sounds recovered as raw electrical signals, we move on to process those.

Recommended cartridge specifications & glossary

Here is what to look for when buying or replacing a cartridge\stylus (targets in **bold**):

- Moving magnet (MM), moving iron (MI, aka *variable reluctance, fluxvalve,* "static magnet"), or moving coil (MC) cartridge (pickup). **MM\MI** have interchangeable styli;

  - Frequency response within + or − (±) **2dB or flatter** <30 to >15,000 (RIAA range);

  - Capacitive load if MM\MI: **within range provided** in\outside preamp, plus cabling;

  - Resistive load if MM\MI: **47kΩ** (47,000 Ohms, all but standardized in most preamps);

  - Compliance Units (cantilever stiffness) >12CU, decent is **18+CU**, lightest are 30CU;

- Tracking force typically **1¼g to 3g maximum** for solid tracking and safe for vinyl;
- Channel separation **>20dB at 1kHz, >15dB elsewhere** (the best are 30dB at 1kHz);
- Channel imbalance **within 2dB** (the best 1dB, but it matters little with preamp gains);
- Vertical Tracking Angle (VTA) \ Stylus Rake Angle (SRA) **15°** (or adjustable in arm);
- Stylus tip **0.3 x 0.7mil** (7.5 x 18µm, microns) elliptical if not a line-contact, diamond.

Starter list of sources online for pickups, styli, info

USA new & replacements suppliers (from the author's most positive experiences): Gary www.thevoiceofmusic.com, Kevin www.kabusa.com, Mike www.esotericsound.com/, Canada - www.canadianastatic.com, Britain - www.pickeringuk.com/styli.html

Disk-ussion – *friendly global online forums [Suggest you have grains of salt handy.]:*

FaceBook: LP & Turntable Enthusiasts, Turntable Talk, Turntable Talk Emporium, Audio Bullshit, VINYL MATTERS RECORD COMMUNITY, The Record Protection Program, Record Collector, Vintage Audio, DIY Audio, Vintage Hi-Fi and Stereo Enthusiasts…

Lenco Heaven – many components other than the turntable - www.lencoheaven.net/

Vinyl Engine – turntable manuals, protractors, product specs - www.vinylengine.com/

WhitePapers on use of subwoofers, and Ambiophonics DSP, etc. – www.filmaker.com/papers.htm

UPDATES this book filmaker.com/papers/UPDATES_RMiller-Better_Sound_of_the_Phonograph.pdf

Add your own below (but scratch off any that prove disreputable)…

While you're at it, is there a turntable, cartridge, or tonearm, or preamp you've taken a hankering to?

Pickup \ preamp \ speakers – a system

We've explored the heart of the phonograph – its stylus – aligned within its cartridge (pickup) "body" with a tonearm and mounted with a "spinner." Individual components comprising what many term a "turntable." But sans preamplifier, power amplifiers, and speakers, the phonograph is silent.

Next in line in the total reproduction system is the preamplifier ("phono stage") that amplifies several thousandths of a volt of pickup output by 100 times to drive a power amplifier. With modern integrated circuits and film capacitors, a state-of-the art preamp has never been more affordable. If you prefer the sound of vacuum tubes, just know that this is due to an order of magnitude higher distortion than solid state, and far less stable over time.

As mentioned previously and in following sections, most suboptimal performance is *not* recorded into the record groove as mastered – they are errors (artifacts) in playback. But most of these can be improved, many dramatically. *You* may make that happen. With the help of decades of the most relevant know-how by the authors and others behind this book.

Beginning this chapter, we explore the interactions and adjustments *after* the turntable, for getting the most from phono reproduction – and thus the most from your records – that otherwise would sound poorer. You might not experience this until you compare results before v. after doing the improvements. One is technically termed the "magnitude transfer function" (MTF, frequency & phase responses) of the whole system, made "flat" to complement the sound's characteristics as engraved in a well-made disk.

A precise (good sounding) overall MTF is an integration of all of the many mechanical and electrical interactions in the phonograph reproduction chain (in the disk from its pressing): between stylus & cantilever, cantilever & tonearm, stylus\cantilever pair & pickup, pickup & arm, pickup & electrical loading, accurate RIAA equalization (EQ, standardized in 1954) by the preamp, also having controls to balance the cartridge's channel-to-channel sensitivities, and having proper monophonic combining to cancel distortion. Flawed performance by any one of these parts and interactions among them wrecks the sound proportionally. But when you have everything working right, you discover how amazingly good the phonograph sounds*!*

Mostly mechanical considerations were discussed in the stylus sections beginning **p10**. A prime takeaway is – considering today's most available replacement styli with spherical & elliptical tips – the better high frequency and lower distortion tracing rests with an elliptical v. its higher wear. But perhaps be on the lookout (and save up?) for the best choice for BOTH sound quality AND wear, though also hardest-to-come-by, a "line-contact," (parabolic, shield-shaped) stylus by whatever tradename.

Often overlooked is *electrically matching ("loading") an MM\MI cartridge at the preamp.* Although critical, it may simply be adding a pair of small capacitors for a proper "C-load." Or easier still, snipping out a pair*!* But because the audio chain is *proportional* to its weaker links, your fine stylus and pickup will not perform their best if the preamp and electrical interface with the cartridge are not implemented to sound as well as designed.

In the realm of active electronics, changes toward a flat MTF playing records you enjoyed before may mean hearing flaws that previously were masked. But more likely you will enjoy well-made records more, a lot more*!* Maker solutions in the next chapters are basic, reflect the technology of simpler times, yet today are "measurably high-end." You won't be bragging about how much you spent, but how little. Proud of the subjectively better results, reflecting your knowledge of sound. Even if you did it yourself*!* [27]

As said in the 2nd *edition Preface* on **p8**, with the analog phonograph, there *are* "user serviceable parts inside." In their heyday built to last, it is not a mystery how to maintain used turntables, pickups, and even needles with low-to-mid mileage, without the financial and environmental impact of replacing & discarding whole subassemblies, as is SOP with digital devices. Optimization is especially possible with a good preamplifier with essential controls. In many cases, external analog preamps have seen the most technological improvement since the 1970s, using high-performance chips and film capacitors, with little remaining to do audio technology-wise. With today's most cost-effective designs, there is no reason for an off-the-shelf state-of-the art preamp to cost more than a couple hundred dollars.

After exploring *how come* phono stages with accurate EQ are needed, there follows a bill-of-materials and *how-to* instructions for such a cost-effective custom phonograph preamplifier that requires medium skills with a soldering pencil and common hand tools. Modification time: ~ 90min once parts are at hand. Later the book presents a bit more difficult tonearm maker project to compliment an already owned but still mechanically sound turntable, or one found used. (Perhaps as a 2nd arm for a 2nd format, up to 16in diameter?) With either project, despite their minimal cost, many will be the years of satisfaction and return on your invested efforts, and enjoyment of the music, as reported by readers of this book's 1st edition.

Even if you do not engage in the maker projects, the information about their set up for optimal performance applies equally to purchased solutions. For any preamp can have the described performance and functionality for best "soundstage" in stereo and lowest distortion in mono by proper C-load, compensating cartridge L\R sensitivities, and proper mono mixing.

Don't warm up the soldering iron just yet *but read on!*

---

[27] As said, manufacturers of integrated turntables might have accounted for mechanical and electrical issues when combining the components of cartridge, arm, & preamp, which the reader is advised to double check. But historically, the very best results typically involve a turntable-only (spinner) and separately mounted tonearm.

Pickup \ preamp \ speakers – a system

"Tuning" the cartridge with the preamp
and a bit more historical perspective, now in the phonograph's electrical era

Named because before the "integrated" audio era it preceded the "control preamplifier" that in turn preceded the power amplifier, a preamplifier aka "preamp" provides the *gain* (voltage amplification) and any filtration required of low level transducers that convert between forms of energy – microphones, tape heads, & phono cartridges – in order to reach audio "line level."[28] For stereo pickups, the conventional stereo reference level is 3.5mv,[29] requiring a nominal gain by the preamp of ~100x (~40dB) at 1kHz. Also providing RIAA "EQ," gain is 10x (20dB) at 20kHz, requiring low distortion, but 1000x (~60dB) at 20Hz, requiring low noise. [The "dB" chapter clarifies gains & levels.] Once a tall order, after Lipschitz & Jung, the maker project later is possible using an integrated circuit (IC) chip.

For the intended tone-color (timbre) of recorded sounds, the preamp must accurately compensate (equalize) for the inverse characteristic used in cutting the master. In 1954, the RIAA standardized this EQ for 45s, 78s, and LPs. It specifies three playback filters: a low frequency *turnover* below 500Hz; a high frequency *roll-off* above 2.1kHz; and a very low frequency (VLF) *rumble shelf* below 50Hz, not to be ignored for listening at performance level even with resonance attended to. Optional are low-pass "scratch" filtering above about 6kHz, and further high-pass for rumble below 20Hz proposed in Europe in 1976, but since withdrawn. So-called "Neumann" boosting at 50kHz is bogus. These parameters characterize the basic design of the phono preamp before considering the controls needed.

Today most consumer phono preamps have no controls. Missing are C-load, gain\balance, and mono mixing. A vintage Apt-Holman preamp has them; a cost-effective Schitt Mani lacks two: individual channel gains and a mono switch, for which one handy enough can derive modifications from the later DIY chapter [steps in this book's Update]. Some enthusiasts need selection of more than one tonearm, and a separate C-load for each.

Archivist\restorer preamps for vintage pre-RIAA disks need selectable or variable settings for any combination of turnover \ roll-off \ rumble shelf choices made by record labels, often in secret, from the start of the electrical era in the mid-1920s for 78s until LP standardization in 1954. The 30yr of chaos, plus prior decades of mechanical trial & error experimentation during the acoustic era, resulted in consumers EQing their replay by ear using the amplifier's single tone control, invented for the purpose. Archivists must use forensic methods to decode early era producers' methods, beginning with the disk speed.[30]

---

[28] For consumer electronics the Institute of High Fidelity manufacturers (IHF), nominal line level is ~300mv rms. For decades, the broadcast industry reference was +8dBm, later +4dBu, or 1.228vrms, ~12dB higher than IHF.

[29] Mid the 1kHz plateau between filter f_3's an octave at either side, RIAA reference velocity is 5cm/s RMS for a mono groove, =3.54cm/s for 45\45° stereo, and allows dynamic peaks (headroom) 12dB higher. On **p63** Shure data show maxima above 70cm/s, likely a pop cresting 20+dB above reference, requiring that much overload margin in preamp design to prevent clipping, or *latching up*, requiring a power cycle. For a pickup sensitivity of ~1mv/cm/s, this pop crests at ±70mv, while a reference tone crests at 8mv (3.54mv RMS each stereo channel).

[30] The EQ characteristics vary with disk speed, which for amateur recordings and most so-called 78s, must be determined first. The author measures the omnipresent hum frequency to correct for non-standard disk speed.

An MM or MI pickup requires pre-amplification to raise its nominal output voltage about 100 times. Moving coil (MC) pickups require an extra 10x (20dB) of very low-noise gain, or a "step up transformer" (SUT). MC were the first "translating devices" (WE), some by Fairchild, Grado, and Ortofon having interchangeable styli. Modern MC styli are not user-changeable, and were not offered by Pickering\Stanton, or by Shure. Boutique suppliers today have made MC a niche product to capitalize upon well-healed audiophiles.

A stylus reproduces the full audio range by its mechanical resonance. Then MM\MI pickups rely on an additional *electrical* resonance to prop up high frequency (HF) response. The "tank" circuit consists of the internal coil inductance in parallel with an external capacitance. Although this C-load is essential to the cartridge design and its performance, it is most commonly a function of the preamp, and so it is treated in this chapter. Also a characteristic of the cartridge due to manufacturing variances but left to the preamp for users to correct is the pickup's inherent channel-to-channel imbalance of up to ~2dB. Or the optimal stereo soundstage and cleanest monophonic sum cannot be achieved.

A stand-alone preamplifier (aka "phono stage") is usually superior to any built into receivers\controllers (few now even have one). A later chapter lists the dozen parts with simple instructions to modify a preassembled preamp board, a $19 import online, resulting in excellent performance for ~$35 total. Yet its design has selectable C-load & R-load, channel balancing, accurate response ±¼dB 30~20kHz, and phase error <4° 200~10kHz.

Affecting everything about sound reproduction quality are imperfect speakers in conjunction with non-ideal listening room acoustics – a subject to fill a sizable library. An audio component auditioned in a typically untreated store showroom will inevitably sound much different in a differently (un)treated home. Norman Pickering wrote that any real audio system might contain an error that by chance happens to offset another, perhaps acoustic one. However, counting on trial & error is a pure gamble, and has been proven online to foster delusional conclusions. Better to attend to each link in the audio chain separately, otherwise any component change may break it, and lead to another wild goose chase to find a different offsetting condition. For some, these fun & games *are* the hobby!

Recording quality can vary; not all releases are made well. For acoustic music, poor acoustics, the main microphone *not* in the best seat in the house; for pop music, multiple microphones acoustically comb-filtering (for a rocky frequency response), monitoring too loud, mixing over-processing, unskilled mastering… A common error is near-field monitor speakers laid on the console meter bridge, where splashes of the sound off the desk surface *comb filter* what the operator hears, causing him\her to "compensate" with EQ and other mixing decisions set in stone that are poor-sounding in subsequent listening environments. [31] The solution is mid-field speakers behind the console, where their sound is *shadowed* by the meter bridge. Mechanical pre-pressing processes are critical, e.g., vinyl composition. But a few of the hurdles each recording, analog or digital, must jump before being played.

Do phonograph records "sound better?" "Warmer than digital?" Professionals recognize digital audio is *technically* better: less noisy; greater dynamic range; no *mono-ing* <150Hz important to spatiality; fewer distortion mechanisms; longer life. It only makes sense that these technologies were developed to be superior. Results then actually *sound* superior if not abused by labels' "volume wars," *normalizing* to the ceiling soft content vying with loud. The egregious policy mastering digital media to raise loudness until audio waveforms are clipped! That's why vinyl can "sound better." It is often more natural, more life-like.

---

[31] Speakers laid on their sides wreck imaging by the speaker's design of HF v. LF drivers not in a vertical line. Frequency response especially around a crossover bakes in errors heard with proper speaker positioning.

Capacitive loading of moving magnet and moving iron pickups

One omission has tainted the reputations of magnetic cartridges for decades – *loading* [Steinfeld 2010].[32] A load *completes* a manufacturer's design of its high impedance MM & MI cartridges. Although critical, it is external, user\installer-supplied, thus beyond control of its maker. In the pickup's *equivalent circuit*[33] **opposite**, the *resistive load* ("R-preamp," "R2" is all but standardized at 47kΩ (47,000 Ohms) and is inside most preamps. However *capacitive load* varies by cartridge. More than 10~20% above or below the manufacturer's specified C-load results in a cartridge being criticized unjustly in many a review as "too brittle" or "too dull\muffled" in HF, or relatively "thin" or "bottom-heavy" in bass. Failure of a reviewer to attend to C-load likely led to false tone color (timbre). Then readers accepted that opinion just as oblivious that no MM\MI pickup can realize its frequency response if reviewers and users have not provided its specified capacitive loading, C-load.

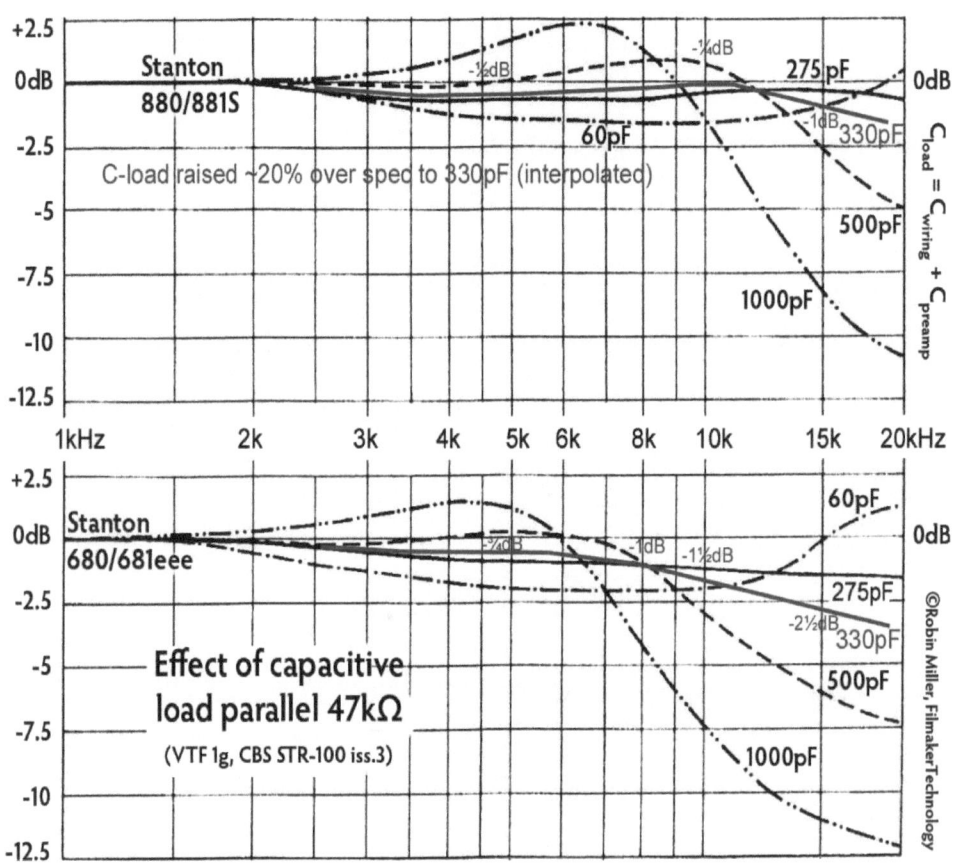

Frequency responses for two cartridges by C-load varying 60~1,000pF. Responsibility of the user\installer, $C_{load} = C_{cable} + C_{preamp} =$ 275pF. Other makes & models vary, creating the need for C_{load} selection. Data courtesy Stanton Magnetics, New York. Adapted from Ballou "Handbook for Sound Engineers: the New Audio Cyclopedia," pub. by Howard W. Sams (1987).

---

[32] Richard Steinfeld *Handbook for Stanton & Pickering Phonograph Cartridges & Styli*, self-published (2010) rsteinbook@sonic.net, a compendium of 100+ products, based on former employee interviews.

[33] All pickups are natively *balanced,* work best into balanced preamps. Common coax wiring *unbalances* them.

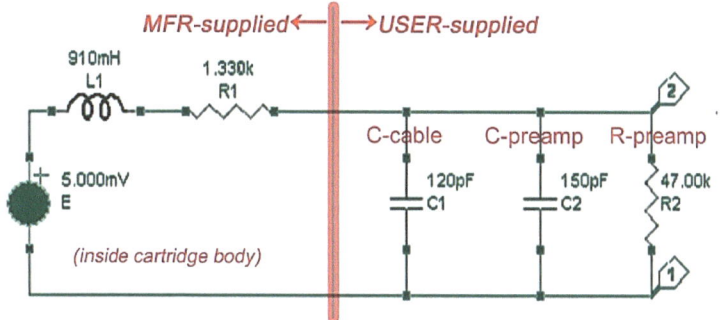

Equivalent circuit of all MM\MI cartridges. Makers instruct users\installers to attend to proper loading, symbolized by C_{cable}, C_{preamp}, & R_{preamp}. $C_{load} = C_{cable} + C_{preamp}$ and varies by cartridge maker. Inside most preamps, R_{preamp} is 47kΩ. Off-spec loading results in replay deemed "too dull" or "too bright." Or randomly compensate errors elsewhere in the system [Pickering 1953].

In the diagram, $C_{load} = C_{cable} + C_{preamp}$. In the hi-fi era of the 1950s, Pickering MM/MI cartridges standardized its capacitive C-load at 275pF (pico Farad, for Michael Faraday) for a high frequency (HF) resonance with each channel's coil inductance and resistance (L1 & R1 in the diagram) to extend the cartridge's HF response. Twisted pair arm wiring plus coax interconnects had capacitance of 25pF distributed per foot along about 5ft, so C_{cable} = 125pF. Then a small a capacitor was manufactured inside the preamp near the input jack in parallel with the 47kΩ resistor, typically C_{preamp} = 150pF. C_{load} then totalled 275pF. *Q.E.D.*

In the chart **opposite**, the C_{load} stories of two higher end of the more than 100 products of musician Norman Pickering and engineer Walter Stanton for home enthusiasts, mastering engineers for quality control while lathing, archivist, and the rigors of broadcasting. Their cartridges require 275pF (pico-Farad) ±10%, or ±20% for a mid-high impedance 880/881. Operating outside these tolerances will have a detrimental effect on sound, as shown.

For both impedance pickups shown, too high a value of C-load creates a peak in the mid-high range, then falls steeply at HF. Too small a C-load moves the peak toward ultrasonic VHF frequencies at a cost to mid~HF. Ignore the designed C-load, the cartridge cannot be deemed complete, and is not flat. To say that proper capacitive loading is important almost goes without saying, yet here we are discussing it precisely because it is so often ignored.

Overall flatness of frequency response curves is an attainable goal, but failing that, an important perceptual truth is that bumps in frequency response are more deleterious to the ear than equal dips. So leveling bumps in favor of broad, shallow dips results in better sound, and determines the choice of 275pF in specifications for these cartridges. However for the 880\881 pickup, might some prefer a gentler mid-HF response of a C-load =330pF?

An enthusiast with a "trained ear" posted in an audiophile forum online these telling impressions of a popular AT 95E that specifies an unusually low C-load =97pF that is hard to achieve with no more than 4ft (1.25m) of the low 20pF/ft coax and zero (no) C-preamp:

C_{wiring} +	C_{preamp} =	$C_{load\ spec}$	Subjective quality of sound
75	47	122	"Impressive, but only initially" [HF dull above a misleading peak]
75	100	175	"Harsh bordering on shrill" [Obnoxious pronounced mid-HF peak]
75	22	**97**	"Cleaner, defined soundstage" [flattest response; least fatigue]

These subjective "results" conjure the Stanton 881 C_{load} response. For the AT loaded with 122pF (cf. the Stanton 881 curve at 500pF) "brightness" is only initially impressive, but "air" at very high frequencies is noticeably attenuated. At the highest specified 175pF

"Tuning" the cartridge with the preamp

(cf. Stanton 1,000pF curve) the AT accentuates the peak and lowers its frequency and sounds harsh. But the lowest $C_{load\ spec}$=**97pF** may be the real AT 95E design value, as it flattens the resonant peak (cf. Stanton curve at 275pF), although it is difficult to achieve. Note that low impedance moving coil (MC) cartridges are not affected by C-load – none is needed. However for more ubiquitous MM/MI cartridges, C-load tuning is important!

But without a means of C-load selection, what if cabling is higher in pF/ft? Or lower? It may be inconvenient to alter C-cable. What if inside is a different C-preamp, or none? On the web are Y-connectors to add to C-load with plug-in capacitance values, but an old engineering adage is: *things break most where they're connected*.

Then what if the cart is a Nagaoka MP-11 specifying from a low of 100 to a high 620pF? What if for playing different types of disks, you switch between headshells with cartridges differing in specified C_{load}? Shure M97 or M35x specify 200~300pF. For its 95E described on the **previous page**, Audio Technica recommend 100~200pF (the lowest 100 is correct). Specifying a range instead of a precise value is a cop-out to curry favor with busy buyers.

For changeable C-load, perhaps make a C_{preamp} switch, as shown below, inside or external to the preamp. Substitute values in this C-selector for the range needed. You can measure the loaded cart's frequency response using a respected, unworn test record & level meter.

What about different arms/turntables switched at the preamp input? To switch-select C_{load} values, drill a ¼in hole, mount a switch, and solder six "caps" (~$10), illustrated next. With pairs of 100, 150, & 250pF and the typical 125pF cabling, you can select 225~475pF.

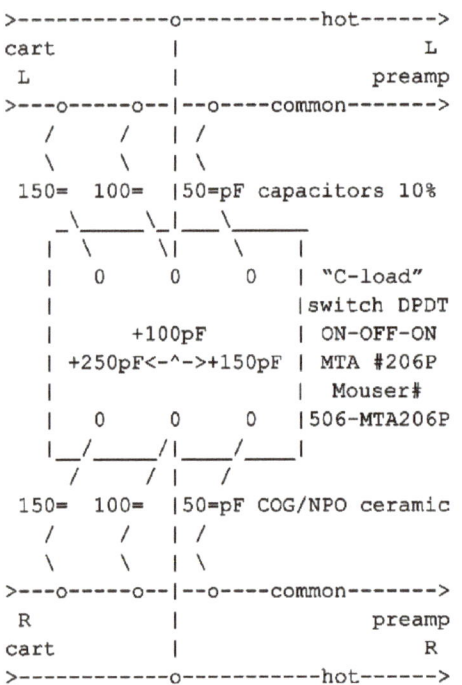

```
>------------o-----------hot------>
cart        |                L
 L          |                preamp
>---o-----o--|--o----common------->
  /      /  | /
  \      \  | \
 150=   100= |50=pF capacitors 10%
   _____|__\_
  | \     \|    \    |
  |  0     0     0   | "C-load"
  |                  |switch DPDT
  |       +100pF     | ON-OFF-ON
  | +250pF<-^->+150pF | MTA #206P
  |                  | Mouser#
  |  0     0     0   |506-MTA206P
  |_/_____/|____/____|
    /    / |   /
  150=  100= |50=pF COG/NPO ceramic
   /     /  | /
   \     \  | \
>---o-----o--|--o----common------->
 R          |                preamp
cart        |                R
>------------o-----------hot------>
```

A $10 DIY selector "tunes" a stereo MM\MI cartridge, **L** in typewriter-schematic, **R** retrofitting yellow disc capacitors ("=" at left, 4 of 6 shown at right) and DPDT (center off) switch to select 100~250pF. Ferrite beads filter radio frequency interference (RFI).

Swapping cables offers another run at the goal: use a $20 pF-level capacitance meter or bridge to measure & tag your cables, and select the nearest fit. Merchandising of audio accessories, especially cables, is imbued with poppycock. *Knowledge is the power to ward off being ripped off.* Consumer audio wires are not mysterious, nor need they be expensive. Maybe free, tangled in your box of spares! Even inexpensive audio "interconnects" work within limits. Capacitance of a cable is proportional to its length: Twice as long, twice the C. Cheapest, usually the very thinnest audio cables measuring 80 pico-Farads (pF) per foot of length are OK if limited to 1½ ft.[34] But their poor shielding may be inadequate at low ~5mv (5milli-volt, 0.005v) cartridge levels v. radio frequency interference (RFI). These are fine as intended, for low impedance line level connections up to 10ft or more.

Mid-priced coaxial cable might have 30pF/ft; the best 20 pF/ft. To increase C_{cable}, pick a less expensive 40 pF/ft cable, or make it longer. To reduce C_{cable}, pick a more expensive 30 or 20 pF/ft coax, or make it shorter. It is not about splurging for highly hyped cable, but to optimize cleverly yet economically the total C-load. Whether 3ft of 40pF/ft cable, 4ft of 30pF/ft cable, or 6ft of 20pF/ft cable, each adds 120pF of C_{cable} toward the grand total C_{load}.

Balancing two channels for mono & stereo groove replay

Ah, *The Golden Mean*, aka *Fibonacci ratio* – the *sweet spot* of visual perception and stock buy & sell timing. Its cousin in audio reproduction is *balance*. Unlike the Golden Mean's 61.8 : 38.2 split, audio balance is strictly 50:50. A small stereo imbalance derails what audiophiles term "soundstage" (audio engineers call it *localization*). Playing monophonic disks using a stereo system, imbalance destroys mono's cleanliness from distortion.

For stereo's conventional equilateral speaker-listener triangle (60° loudspeakers), a difference in level between channels of only 15dB pans an auditory event fully to one side or the other, emanating from the speaker on that side alone [Theile 2001]. In the middle, it takes little difference to move the image quickly off-center, skewing the soundstage non-linearly. This is perceived to leave a "hole in the middle," shoving farther out voices and instruments recorded only slightly left or right. If a stereo recording is reproduced only slightly off balance, say by only a couple dB of the 15, the entire image will shift quickly toward that speaker, leaving void through the center and toward the other side. [35]

Assuming a well-made recording (a big assumption), every component of an audio system can introduce imbalance due to manufacturing tolerances in electronic gain, loss, or transducer sensitivity. Some errors cancel, others add to a worst case. And these may degrade over time. The "balance" control was invented to compensate for a system's sum of imbalance errors, adjusted by ear, so often needing tweaking. Even with the rest of the system precisely balanced, playing with a pickup cartridge that just meets its specified 2dB sensitivity imbalance will throw off its imaging if not corrected in the phono preamplifier.

Cartridge manufacturers did not intend this output sensitivity variation go uncorrected in the preamp, just as tape machine makers provided channel gain controls for variations in the playback tape head-stacks. Integrated amplifier/receiver makers fell back on the overall balance control to save the cost of a few parts, or user complaints of too many fiddly bits.

Playing vintage monaural vinyl is affected differently, but critically. The mono record groove is usually modulated laterally. Cylinders and early disks were recorded vertically, "hill-and-dale," but suffer higher distortion. The vertical "drive" – the groove forcing the stylus up but using the cantilever's springiness for its descent – is mechanically "single-

---

[34] Multiplying to the needed 120pF =80pF/ft x 1½ft, to the preamp inside\adjacent to the turntable base (plinth).

[35] 60° speaker stereo is half a typical recording angle of 120°. Crosstalk cancellation (virtual headphone listening) reproduces the full 120° "soundstage" without stereo's hole-in-the-middle. The author engineered Ambiophonics' RACE and PanAmbiophone mic – www.ambiophonics.org & www.filmaker.com/papers.htm .

ended," so is susceptible is to added even-order harmonics. Lateral groove walls drive the stylus both left and right by the groove, "push-pull," so does not add even-order distortion. Most sounds consist of either all harmonics OR odd only. Just as clipping adds odd order harmonics that alter tone color, adding even harmonics fouls the *timbre* of the original sounds.[36] Timbre (tone color) is the main reason harmonic distortion should be minimized.

Early stereo in a single groove used lateral motion for the Left channel and vertical for the Right, which contained more distortion and surface noise, and therefore sounded comparatively inferior to the Left. (Early stereo radio *simulcast* one channel on AM, the other on FM.) Stereo caught on after L & R was assigned to its own 45° wall of the V-shaped groove, each with modulation equal parts lateral & vertical, as in the following vector oscillographs. L & R channels shared distortion equally. With no perceivable difference, the balanced distortion was acceptable, although it was and is still there.

These artifacts are mostly vertical motion, compared in the next vectorscope images.

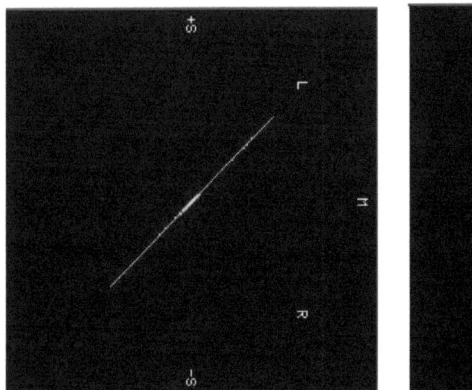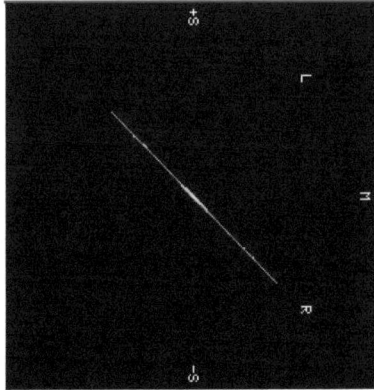

Vectorscope images illustrate stylus motion viewed from the front. **Above L,** signal only on the "L" channel and groove wall, as labeled on-screen. **R,** signal only on the "R" channel and groove wall.

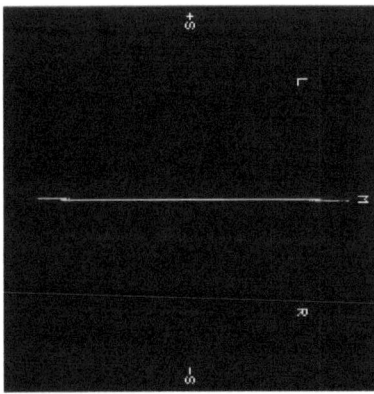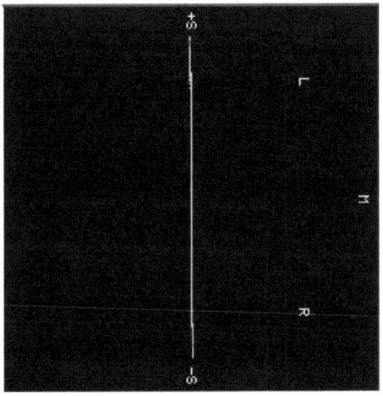

L, lateral-only motion (mono, M). **R,** vertical-only (out-of-phase, "S" intentional in stereo, but distortion in mono). Any way rightward are positive-going signals in-phase for both channels.

---

[36] Stringed instruments (violins, guitars, pianos, human voice) and conical-bore saxophones generate even & odd *harmonics* [Sauveur 1700, Rayleigh 1877]. Clarinets & trumpets produce *only odd* harmonics. No acoustic instrument creates even-order harmonics alone (excepting synthesizers, including Hammond drawbar organs).

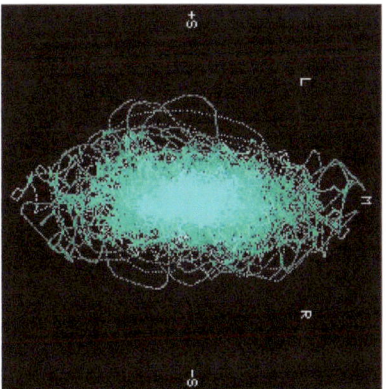 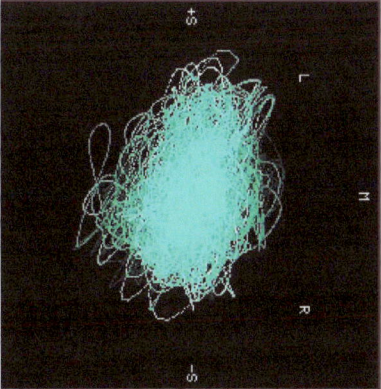

L-Highly correlated stereo (mostly mono).   ***R***-highly *un*correlated stereo, "phasey" and at high levels more likely to jump the groove. Both show how a centered voice as at left, but if distorted with sibilance spit added by "pinch effect" for that instant looks like right.

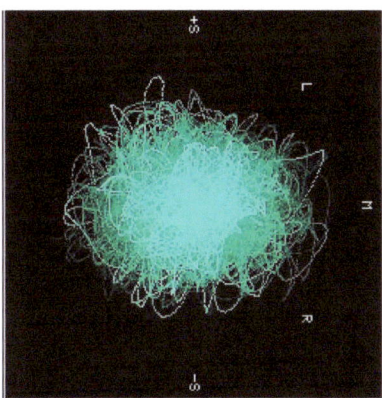 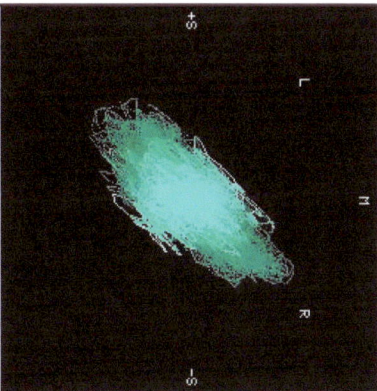

L, well balanced, spatial stereo of a grand piano in a concert hall. ***R,*** same piano softer, accompanying a solo violin standing at right. This ferocious groove and stylus activity is purely mechanical – unlike those nonchalantly inconvenienced electrons and digital bits. Steinfeld marvels that "The phonograph sounds better than it has any right to!"

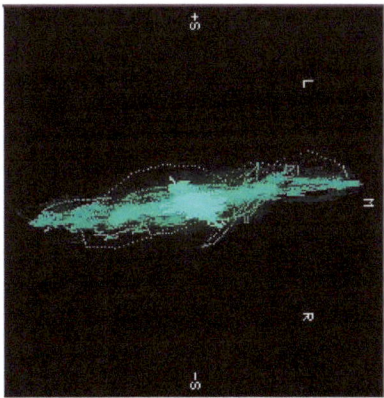 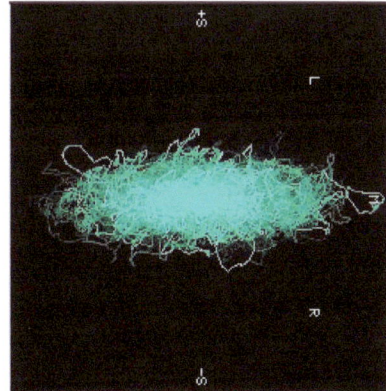

L, after channel balancing using a test record, this kick drum, likely intended to be panned center, shows a recorded imbalance of 1.2dB left of center. ***R,*** a mono disk played with a stereo cartridge contains unrecorded vertical distortion that should be nulled to appear as Lateral-only (mono screenshot on the prev. page). 2nd harmonic pinch & poid effects are out-of-polarity in stereo, disembodying transients & HF as spitting sounds flashing around the room outside the speakers*!*

"Tuning" the cartridge with the preamp

From the initial vector figures on **p78**, the cartridge's output signals polarity for *a rising motion are positive-going on the Left channel but negative-going on the Right*. Hence the 2nd harmonic distortion produced is opposed in polarity! 180° out-of-phase. Adding pinch\poid effect distortion causes the phony partials (only, not the fundamental) of voices and instruments to appear instantaneous beyond the speakers, flashing randomly around the listening room. These errant spitting sounds of vocal sibilants, buzzing brass, and dirt pops & percussive transient ricochets are ghostly, harmonics dismembered from the root sound, and defocusing the intended images of voices and instruments throughout the soundstage, most importantly affect soloists near the center. These spurts of hash are most likely to appear toward the inner groove region during a climactic cut on that side.

For best mono, undesirable vertical artifacts should be nulled simply by summing L+R *after* signals from the groove are in balance, by separate preamp gain controls adjusted upon installation of the cartridge. *These artifacts remain and are heard in stereo.* The simplest (but not best) way to cancel unwanted vertical components playing monophonic records is to parallel (short-circuit) the L+ and R+ outputs of the cartridge, usually at the cartridge, or by a switch at the preamp input. However, the channel sensitivities of many a stereo cartridge might be different by as much as 2dB, as published in many manufacturers' specifications. When channels differing by 2dB are shorted together, the reduction of bogus vertical hash is only −11.7dB. [37] As a change in level of 10dB, softer or louder, is perceived as halving or doubling of volume, the distortion would remain half as audible as before – still poor sound. Balancing cartridge outputs using preamp gain controls nulls vertical tracing distortion by far more, rendering vertical distortions inaudible.

It may be the cartridge is slightly rotated in the headshell or arm, or the disk has been cut with a slight error in balance, depicted in the bottom left capture on p81. An imbalance in the record groove might have been caused by a mastering error, such as when mono new and reissued LPs after 1958 increasingly were cut on stereo lathes. Then a total imbalance of 2½dB and merely paralleled without balance correction reduces vertical artifacts only 9.5dB, still very audible. For pinch distortion reaching 10%+, simply paralleling cartridge outputs for monaural reduces it to 3.3% or more 2nd harmonic distortion – not exactly high fidelity. (Note that the pinch distortion when reproducing in stereo remains up to 10%+!)

Because this issue is so often ignored, it bears repeating that it is best for mono to mix the L & R signals *after* or within a stage of pre-amplification, where L or R cartridge errors can be trimmed to be in balance. Then summing L & R, vertical artifacts are nulled. In the real world held to ±¼dB difference before summing, the reduction of distortion and vertical noise would be 97.1%, attenuating artifacts by 31dB, resulting in the worst case of less than 0.29% maximum distortion contribution of *both pinch & poid-like vertical artifacts.* This fits well the definition of hi-fi if other distortions are similarly optimized. No more spitting sibilants, not in mono anyway. This case of a 1,000+% improvement over paralleling cartridge outputs is easily achieved by a gain trim-pot in one channel, or by switching gain in ½dB increments to a total of ±2dB, as in the modified RIAA preamp in the next chapter.

Do NOT simply "Y" the preamp outputs, causing severe distortion, instability, and latching-up (requiring a power cycle). With solid state low impedance outputs, each channel would short-circuit the other! Instead use "build-out resistors in the DIY chapter.

While a stereo cartridge, having vertical compliance as well as horizontal, is the safer way to play monophonic records, it reproduces unwanted vertical junk that isn't recorded

---

[37] An exception is the Decca-London cartridge with lateral & vertical coils (M-S), so paralleling channels cancels vertical distortion artifacts precisely. However, its stereo "soundstage" balance cannot be optimized by gain trim.

content, but artifacts to be minimized. The spitting sounds of vocal sibilants and other HF artifacts above the stylus' "cutoff frequency f_3," when a spherical or fat (0.4mil) elliptical stylus cannot fit high frequency cornering, especially in inner grooves, as on **p22**. Or losing contact with one groove wall due to too little tracking force for a low compliance stylus. Too little or too much anti-skating force. Optimized by alignment, these tracing distortions augment those of angular tracking errors of the tone arm and pickup [Yamamoto].

The bottom right vector image on **p79** shows LF vertical artifacts are far lower than HF. Some preamps provide a switch or variable frequency mixing L & R channels <150Hz to cancel those, duplicating monauralizing during mastering that left no LF vertical content (S), along with vertical spinner rumble that might have reached the audible range.

Playing digital is *so* much easier, yet analog "vinyl" is in revival. In part because analog electronics such as a phono preamplifier have evolved to where they are no longer a weak link. Limiting both digital and analog audio are loudspeakers and listening space acoustic treatment that require the most attention and financial resources. The preamp modifications in the next section by your humble audio engineer and skinflint will be comparable to many, even though you made it! No matter what others would have you *believe,* often with no proof, the results from your cartridge and disk collection from following this paper will be hard to beat. For a change, enthusiasts can be satisfied and maybe even brag about *how little* they spent! And how no more than a few hours effort paid off in years of enjoyment.

We've only "scratched the surface" of disk history!

Briefly, following Edison's cylinders (1877), after Berliner improved distribution with better and cheaper flat disks (1887), by the turn of the 20th century the Victrola succeeded the player piano in home entertainment popularity. Electronic recording began in the '20s, after a hiatus for the World War, using vacuum tubes of Fleming and de Forest (both 1904) and their application development by Armstrong (especially for radio). After another hiatus for WWII came magnetic tape (1947), micro-groove LPs (1948) and 45rpm singles (1949).

Stereo on disk was first realized in 1952, the Cook double-spiral (dual mono) disk played with a forked arm carrying two pickups. [Paper at www.filmaker.com/papers.htm]. The author heard his first stereo from a Cook LP he recalls as scary stories told only in sound effects. In 1958 45°\45° single groove stereo reappeared, credit in 1932 both WE\Bell Labs and Blumlein at EMI, each unaware of the others work. Significant advances were variable pitch\depth and heated cutters by Scully, Grampian, Neumann and others beginning in the 1960s. Double speed mastering put icing on the cake, but could not be checked on the fly.

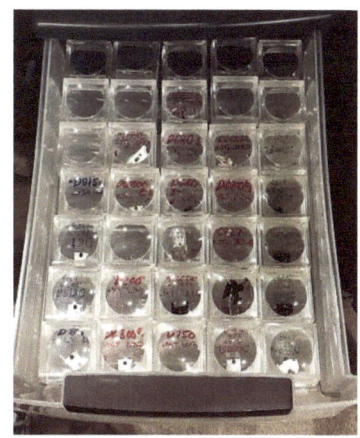

We've only "scratched the surface" of disk history!

Above is a drawer of 35 "bug boxes" with 3x magnifier tops to organize styli by groove and tip sizes, MI or MM, and VTF. And two of this book's "steampunk" 12in (305mm) tonearms separately aligned for microgroove (fine groove) LP and wide\coarse-groove SP. Overhanging the platter is a broadcast 16in *electrical transcription* (ET), wide-groove but 33⅓rpm, also used for early talkies as Western Electric's "Vitaphone" (~1925). Compared to only 3½min max for a 10in 78rpm or 4½min for a 12in, a 16in ET side lasted 20min. Modulated either laterally or vertically (mono). Labeled to start either outside or inside.

The 16in ET's capacity established 15min as the duration of most programs in The Golden Age of Radio before Bing Crosby first used magnetic tape (1949). For 30 or 60min programs, two turntables were used, the 1st disk beginning outside for a loud introduction, segueing to the inside of the 2nd disk, etc., ending outside for the program's louder climax. As disk quality improved, disks were segued conventionally, until ETs were microgroove holding 30min+ per 16in side. Supposed to be destroyed after two airings, those saved are collectable today. Another grooved format containing recorded history exclusively.

From the 1940s into the '60s, for commercials, jingles, and private recitals, radio stations also could record in-house their own one-off ETs, called "instantaneous acetates" (playable right away), actually lacquer coated aluminum using a mid-quality lathe. At local stations around the USA, popular semi-pro Presto disk recorders with limited sound quality of only 100~6kHz were operated by a member of the staff who had other primary duties. Similarly, local Victrola stores offered recording services to church choirs and music conservatories.

Also using lacquer media, professional mastering for mass duplication used precision record lathes with heated styli and advance-reading tape-heads for automatic control of groove depth & pitch (spacing, aka automatic margin control, AMC). Like avoiding broadcast over-modulation, limiters keep cutter chisels from excursions into adjacent grooves, and protect cutter heads from overheating, but with imperceivable effect on audio.

Local stations, the networks, and stores made disk recordings directly of live performers. Typically using a single microphone, musicians, actors, or Little Suzy during or repeated after a live "recital" on-air recorded an entire song or disk side non-stop – any mistake meant redoing the side. This procedure applied to all recording 1925~1950 prior to wide use of magnetic tape. Even after wide-spread use of tape as an intermediate convenience for recording indirectly to disk, audiophile "direct-to-disk" albums (typical 5~6 songs per side lasting 20+min) have pristine quality from one less generation of degradation in signal quality in order to lower distortion (odd-order harmonics generated by magnetic hysteresis).

Mastering from tape still required changing in a few seconds between cuts (bands) the next one's level, limiter threshold, EQ, even tape head azimuth, as mixes from different studios, or different days, were spliced to form a continuous side for cutting. Before going to the pressing plant for a "test" pressing, the engineer cut a one-off "acetate" ("dub") for the album producer(s) imprimatur, dubs for talent, and some mailed to stations as advance promotional copies. Later, in the new "fix it in the mix" mentality, "mastering" became a new job, a new intermediate step for CDs, and now for new vinyl releases and rereleases.

With all its moving parts, "It's a wonder that records sound as good as they do!" marvels Richard Steinfeld, author of "The Handbook for Stanton-Pickering Phonograph Cartridges and Styli." That sentiment was first said by Harvard researchers Pierce & Hunt in 1938. And it echoes consumer response even in the acoustic era before the mid-1920s – the *wow factor* hearing recorded music for the first time – after his cylinders, an Edison disk (1877). In his 1915 Diamond Disk Challenge, audiences could not tell the difference between live musicians on stage v. the recording of them. It was a trick: Edison hired singers who could imitate the strange sound of the phonograph, not the other way around!

The solid state era of audio entered with the transistor, then hundreds of transistors in an integrated circuit chip. By the 1980s, solid state (SS) empowered ultrasonic sampled digital recording, and analog audio entered its *long tail* technically, in consumer sentiment, and economically. Improvements have become incremental, if even discernible. To remain relevant and profitable, novel sound-seeking labels opt for distortion coloration by digital signal processing (DSP), easily overdone, making the artificial sound which for many consumers has become "acquired taste." Yet far less processed vinyl is appreciated by a natural sound seeking younger market. And records are *a main source* of digital content.

After dwindling in the US to a handful of pressing plants of half-century-old machinery, new plants are being built that can squeeze a 5oz "biscuit" of vinyl into an LP every 30s! Kiosks at Barnes & Noble stores that began with a 6ft long shelf now occupy an entire aisle. Mom & Pop used vinyl stores serve collectors and trend-worshipping teens who enjoy larger album art, and romance placing a rock in a ½-mile long rut to hear music.

128ft of shelving displays new & used vinyl at a store in Mesa AZ as "vinyl" sales rise again!

L: At Marin CA Barnes & Noble, Lia & Ada buy LPs – Beatles for grandsons including mp3 CD-R; jazz standards for themselves. *R:* Gen-Z grandsons mug scratching with the steampunk tonearm made with GramPa.

Evolution of grooved disks: 78 rpm to 33⅓; SP to LP

Until the arrival of magnetic tape in 1947, lacquer replaced wax for mastering disks. Not having to wait a week for a 78 pressed in shellac, immediately playable were "instantaneous acetates," or "electrical transcriptions" (ET), recordable on portable lathes. It was the way radio stations recorded programs and commercials for later air. Syndicators mailed thousands of mass-pressed 16in ETs to stations, and overseas for the Armed Forces Radio Network. With a 2mil (50μm) chisel at 33⅓rpm, the frequency response of these ETs was preserved by a 16in disk's much higher linear groove speed. Indistinguishable from live in fidelity on AM, the FCC required stations in the *Golden Age of Radio* 1930~62 to announce the program was "transcribed" (recorded) so as not to mislead the public, for fear they'd storm the station, expecting to meet celebrities! (Today, what isn't recorded?)

Consumer 78rpm SP disks contained 3½~4½ minutes a side; 16in ETs could hold 22min! Also the duration of 35mm film reels and 16in ETs for Western Electric Vitaphone sound, synchronized to the projector gear-driven to a 16in turntable. **Below**, a 16in ET shown next to other sizes, and overleaf on a popular RCA broadcast turntable with SP & LP tonearms. Microgroove 16in ETs holding ½hr+ lasted until long after introduction of the LP in 1948.

CW from L: A 16in ET with 15min of radio content (and inside start); 45rpm single introduced by RCA in 1949 (7min max for extended play EP); ubiquitous 12in LP (25+min/side) invented by Columbia 1948; Pathé 10in single 80rpm hill-and-dale (vertical, 3½ min per side) with U-shaped groove for a tip 8mil (200μm) wide!

 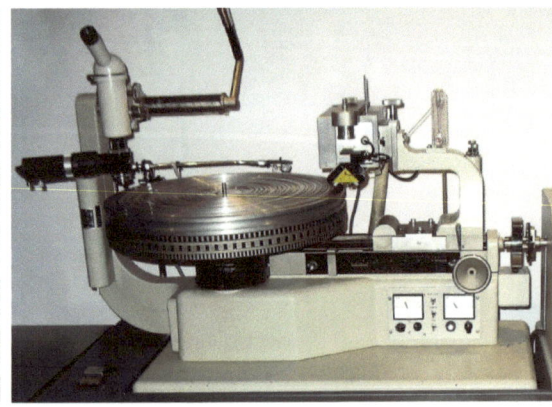

L: Presto portable disk recorder popular with radio stations (electronics in a separate similar-size case).
R: Neumann mastering lathe with stereo cutter, inspection microscope, and work-light arm, electronics in a rack, not shown, plus a master tape player. For QC, the 12in arm at back has a Stanton cartridge.

Broadcasters used two or three 16-inch turntables, like this author-restored 1960 RCA 3-speed puck-drive. The 12in arm at the right is for 7, 10 & 12in disks; the 12in arm at back for 16in ETs, all four sizes shown piled on the platter. The 16in format accommodated 15min *transcribed (not live)* programs, or 30min and hour-long shows by *segueing* between two turntables. Interchangeable styli accommodated all formats.

Records "equalized" for lower noise, no groove hopping

From 1926 as new electrical recordings needed compatibility with millions of acoustic wind-ups, hundreds of recording characteristics evolved by trial & error, a moving target as wind-ups were still experimenting in the art of the "sound box" (pickup). To thwart competitors, record labels were secretive about their formulae. Increasing numbers of radio-phonograph consoles featured a variable "tone control" (treble only), adjusted by ear to listener taste in variable acoustics. Broadcast turntables had a 3-way switch selector for guessing by ear about "decoding" the recording. With both consumer and broadcaster controls plus ears in the audio path, listeners suffered a wild range of sound quality.

Consumer & broadcasters complaints won standardization in 1954 by the Recording Industries Assn of America, adopting RCA's *New Orthophonic* characteristic [Moyer 1953]. Using three simple filters, along with their inverse filters in the mastering lab, the implementation was inexpensive enough for home radio-phono sets coming into popularity. A compromise close to existing standards (within 2dB of AES, NAB, LP, & FFRR curves), RIAA was adopted almost immediately in the US. In Europe it took 10yr to be universal. The inverse RIAA curve for mastering is shown in the recording characteristic **overleaf**.[38]

---

[38] Electro-magnetic record cutter heads and replay cartridges are devices characterized as "constant velocity," not constant amplitude like most of an audio system. This means groove swings halve every octave increase in

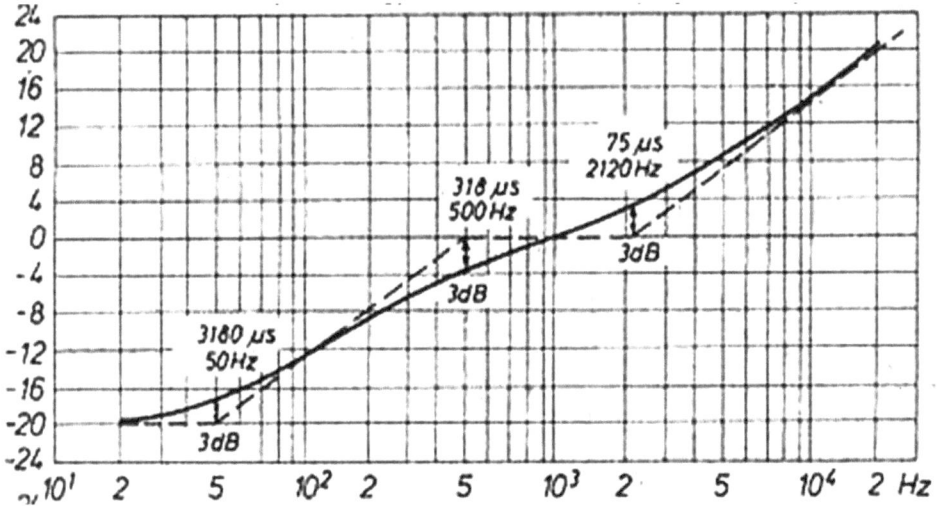

A Bode Plot (straight line) of the *inverse* RIAA curve for cutting mono and from 1958 stereo masters. Consumers play using its mirror image, the RIAA curve. Break points (f_3's) are at 50, 500, and 2120Hz.

The graph above is the *inverse RIAA* [1954] standard for "encoding" all vinyl records. Mastering using a precision record lathe engraved (cut) the spiral groove in lacquer coated aluminum a full side of audio content (initially live, then from magnetic tape). Circuitry involves three simple 1st-order filters using common resistors and capacitors. Reducing LF steadily below 500Hz, ending at 50Hz, lessens over-modulating into the adjacent groove. Rising HF above 2,120Hz while cutting reduces scratch & dirt noise on playback. Not recognized (or shown) upon replay-only are a European IEC 20Hz filter for *infrasonics* [Howard], nor the so-called "Neumann curve" some believe in error is required at 50kHz.

On replay, the "phono stage" (preamplifier) "decodes" using a mirror image of the above inverse curve, after correct C-loading of the cartridge, so the recording's frequency & phase responses are again "flat." At the end of this round trip, tone color is not altered, and the restored phase does not smear sonic images or defocus transients. Now this "RIAA curve" attenuates HF at the rate of 6dB/octave, reducing groove noise, clicks & pops at the same rate. However, LF boosted by 6dB/octave risks high amplification of rumble & hum.

Opposite are typical energy spectra of two recorded sounds. The top shows a piano's mild harmonics attenuating steeply at ~12dB per octave. In the bottom graph, a brass band blaring spectrum descends 5dB per octave, just above the RIAA tilt up of 6dB for HF. So RIAA above 2kHz has "room" for most natural acoustic content without overmodulating.

To avoid problems using rugged juke box and mass consumer styli, many high-HF energy 70s pop LPs were cutoff above ~8kHz [measured by the author and confirmed by two of that era's mastering engineers]. Otherwise the low-compliance cantilevers and pinch-prone spherical tips were unable to trace high levels of high frequencies, generating copious "sibilance" distortion and damaging grooves including HF "erasure" in a few plays.

---

frequency. Otherwise LF would swing widely, while HF would diminish to be lost in noise. So disk "EQ" (actually encoding & decoding) converts the constant amplitude recordings to approximate constant amplitude grooves, and back again. But if this EQ tilted the 10 octaves the required 6B/oct, a flat signal would span 60dB, which is impractical for amplification to fit between overload & noise. So *New Orthophonic* and successor *RIAA* EQ incorporate a plateau in the middle to reduce the span to a doable 40dB, up and down by 20dB referring to 1kHz. The enclode & decode curves fit the dashed Bode plot between the standard's three filter cutoff (f_3) frequencies.

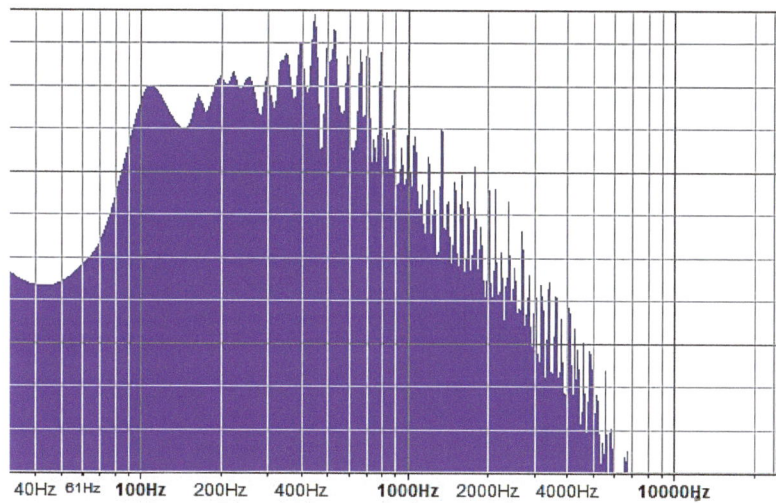

Above: Energy distribution (6dB/div. vertical) of a piano recording is nil above ~7kHz.
Below: An acoustic 17-piece big band extends HF to 19kHz. In grooves encountered by the stylus, inverse RIAA has boosted HF to velocities in the topmost circles on **p63**.

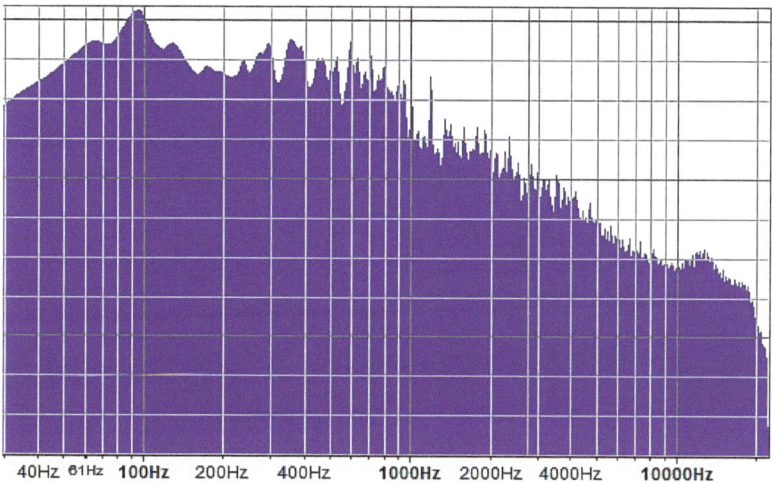

Catering to lowest common denominator users, this practice was a disservice to the music in many fine pop albums in a golden age of rock 'n' roll. Acoustic classical & jazz samples were all full range, expecting these would be played on better systems and using finer styli. By the 1980s, pop mastering was full range as well, perhaps anticipating full-range CDs.

Most stereo LPs are mastered to "monauralize" (mix) low frequencies below ~150Hz.[39] This practice reduces groove hopping, and vertical turntable rumble if mixed again in the play preamp. But it also degrades human binaural perception of interaural time difference (ITD) cues down to 90Hz [Bose, 45Hz the author] that are important for conveying stereo's realistic spatiality. [See author papers "LEV and the Money Seat of Stereo," "Physiological

---

[39] In mastering, mixing $L_{VLF}+R_{VLF}$ redistributes 150~250Hz energy from vertical to lateral. Reduces city studio subway noise, permits easier sanding of the stamper (a "negative," where grooves are ridges), and acoustically couples woofers for stronger bass [Krepps interview by the author 2017]. Note: *When playing mono disks with a stereo pickup, removing vertical spinner rumble, if needed, calls for also mixing <150Hz in the play preamplifier.*

and content considerations for a second low frequency channel for bass management, subwoofers, and LFE," and "Subwoofer Camp" at http://www.filmaker.com/papers.htm .]

The so-called "Neumann Curve" is ill-advised. Proponents believe it necessary to add to RIAA a +6dB/octave rise at 50kHz to compensate low-pass filtering during mastering that protects the cutter head. However, *Neumann's cutting amplifier had a –12dB/octave 2-pole filter* not only to flatten but to turn downward any ultrasonic drive, and earlier at 35kHz! At the accepted audible HF limit of 20kHz, the effect of a 2-pole filter is a negligibly small fraction of a dB. So even if any VHF audio could have made it through disk-making and cartridge replay, it along with any bogus "correction" would be inaudible – except for increased pops from radio frequency interference *even when a disk is not playing*!

While championing standardization and higher quality for both its consumer and broadcast businesses, RCA Victor debuted with much fanfare, but then silently dropped "Dynagroove" to counteract the distortions inherent in the run-of-the-mill spherical needles. RCA inversely pre-distorted the audio in the groove to cancel replay artifacts. But it ruined the sound for sophisticated audiofans using low-distortion elliptical and line-contact styli.

Now exceeding digital CD and download markets, the RIAA reports vinyl sales in 2021 in stores topped $1B (not including unreported private sales). Used vinyl shops respond to music-lovers who like the larger album format for visuals and liner notes. Seniors wax nostalgic for "warm vinyl sound" (dynamically uncompressed), despite prices of $30~40 indicating the trend in demand, and by inference the demand for record playing equipment. (But please consider not risking your collection or your sensibilities by using a $99 player.)

The Library of Congress has a large repository of groove media. The White House has a sizable collection, representing the varying tastes of past presidents. Individual enthusiasts have records in the thousands, as did every local radio station of their target audience's favorites. [40] Historic content mostly not available in digital form is on the longest-lived medium even for data, the shellac disk. Expected to last 200+yr, just don't drop it, or look at it wrong! In time, vinyl was added to shellac for resistance to breakage and better sound.

STANTON	TRACKING FORCE	ORIGINAL STYLUS COLOR	All are products of Stanton Magnetics				Stanton Number	Tip Size & Configuration
500 BROADCAST SERIES	1 to 2 grams 2 to 5 grams 3 to 7 grams 3 to 7 grams	White, Gold ● White, Red ● White, Aqua ● White, Blue ●	** ***		820-DEE 820-DE 820-D7AL 820-D3	27.50 23.50 15.50 24.50	D5100EE & D-5, D-51 ☆ D5100E ☆ D5107AL ☆ D5127 ☆	.3 X .7 Ellipse .4 X .7 Ellipse .7 mil Conical 2.7 mil Conical
600 BROADCAST STANDARD H. P. SERIES	1 to 2 grams 1½ to 3 gr. 2 to 4 grams 3 to 7 grams	Black, 600 Black, 600 Black, 600 Black, 600			821-DEE 821-DE 821-D7A 821-D3	32.50 29.50 24.50 21.50	D6003EE & D-6,61,62 ☆ D6004E ☆ D6071A ☆ D6027 ☆	.3 X .7 Ellipse .4 X .7 Ellipse .7 mil Conical 2.7 mil Conical
681 CALIBRATION STANDARD SERIES WITH BRUSH	3/4 to 1½ gr. 3/4 to 1½ gr. 1½ to 3 gr. 2 to 7 grams	Black, EEE Black, Silv. ● Black, Silv. ● Black, Blue ●			822-DEEE 822-DEE 822-D7A 822-D3	49.50 44.50 36.50 36.50	D6800EEE ☆ D6800EE ☆ D6807A ☆ D6827 ☆	Stereohedron .2 X .7 Ellipse .7 mil Conical 2.7 mil Conical
680 SERIES STEREO STANDARD	2 to 5 gr. 3/4 to 1½ 2 to 5 gr.	With Brush Minus Brush	* ***		824-DSL 824-DE 824-DEL	49.50 36.50 34.50	D6800SL ☆ D680,D-65 ☆ D6800EL ☆	Stereohedron .3 X.7 Ellipt. .4 X.7 Ellipt.
881S PROFESSIONAL CALIBRATION STANDARD	3/4 to 1½	White			825-DEV	84.50 ☆	D80,81,81S	Stereohedron
L747S L737S L737E L727E L720EE	3/4 to 1½ 3/4 to 1½ 3/4 to 1½ 3/4 to 1½ 3/4 to 1½	Clear Clear Clear Clear Black	***		826-DEZ 826-DEX 826-DEV 826-DEL 826-DE	49.50 42.50 36.50 29.50 22.50	D74S Stereohedron ☆ D73S Stereohedron ☆ D73E .3X.7 Elliptical ☆ D72E .3X.7 Elliptical ☆ D71EE .4X.7 Elliptical ☆	

P21 from the last of 30yr of Pfanstiehl catalogs listing thousands of styli for all cartridge brands. Their labels (5th column) indicate original equipment mfr (OEM) if three leading digits, generic replacements if given a leading "4."

---

[40] These collections maintain the equipment to play properly grooved media, from: Pathé hill-and-dale (vertical) 80rpm shellacs (8mil spherical stylus); shellac 78.26rpm SP (standard-play) acoustically recorded until 1926 and electrically after; late 1950s vinyl 78s; SP & microgroove ETs 8~16in; 7 and 12in 45rpm singles from 1949; and from 1948, 33⅓rpm 10in & 12in mono LPs. From 1958, stereo at 33⅓rpm and "audiophile" releases at 45rpm.

The author's "archiving" preamp has balanced inputs & outputs, two arm selection, lateral or vertical mono, five EQ presets (indicated by color LED), variable LF turnover, variable HF rolloff, variable VLF rumble shelf, C-load selection, mono mixing for stereo pickup, variable VLF vertical cancellation. And sounds fantastic!

Above: In mono vacuum tube era, a popular H.H.Scott 121C preamp had LF "turnover" and HF "rolloff" selectors for 78s & LPs.

Above: A 2-speed belt-drive player has a *linear tonearm drive* for zero tracking error.

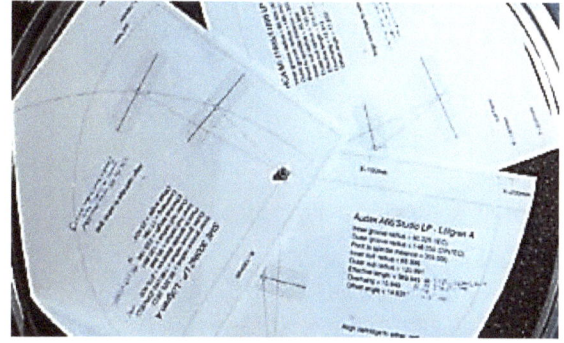

Left: Paper protractors to orient the cantilever tangent to the groove are free to download online. Mirror-surface protractors allow greater accuracy and easier use, such as the author's design on **p60**.

    So far, this book is for those who would benefit from understanding the complexities of the phonograph. A compact reference while implementing that knowledge for better sound. Cleaner audio for better enjoyment of recorded music. Tone color is affected by tracing & tracking *non-linearities*. By stylus tip shape, cantilever alignment, and compliance resonating with tonearm mass. And by load capacitance and the filter accuracy of the preamp. All must be tuned together to convey transparently that intended by its performers and producers. Ultimately for rendering clarity to the window of enjoyment by listeners.

Evolution of grooved disks: 78 rpm to 33⅓; SP to LP

Measuring sound (science alert)

Upcoming sections will be more easily understandable given this primer\refresher about audio level & gain\loss measurement. At the cost of a few pages of reading, it teaches a few tricks for calculating decibels (dB) in your head. While doing away with the cumbersome logarithm math, it gives readers a quantifiable context for what we hear. A perspective that avoids making the impossibly perfect the enemy of the quite good enough.

More precisely, how our brain interprets (perceives) the "auditory events" relayed to it by our ears with each passing microsecond! For as important as are frequency response, noise, and preamp gain, most important is *psycho-acoustic perception*. Both the *qualifiers* of hearing as well as the *quantifiers* of audio performance are based on measurements in dB.

The decibel is also an example of the essence of engineering – aka applied science. Unlike physicists or mathematicians, engineers are not *perfectionists*. Rather, we are *precisionists*. Given the degree of precision desired, the engineer will tell you what you can get away with. How much less steel in a bridge is unsafe; how much more is needless cost. Engineers apply mathematics and the physical sciences in useful, practical ways. And balance performance within limits of size, weight, power consumption, ease of use, cost, etc. For the DIY projects in this book, an engineer (me) found solutions for which their low cost belies fine performance. But instead of licensing them for millions, I practically give them away free in this book! [41]

Edison patented the reproducing phonograph in both cylinder and disk form in 1877. Rather than a calculating engineer, he was a tinkerer who invented by the protracted process of trial & error. That same year, a real scientist, Nobel prize-winning physicist John William Strutt (Lord Rayleigh) published his *The Theory of Sound* taking a purely mechanical approach, then in 1894 to passive electrical (no *valves* yet). Audio as a discipline really got going in the 1920s with the first *active* electronic component, the vacuum tube, which made possible long-distance telephone, radio, and the electrical recording of disks. Bell Labs invented the decibel (1/10 Bel) in 1923, naming it for Alexander Graham Bell. By the 1930s, several audio engineers devised stereo, and foresaw elliptical and line-contact styli.

One decibel (1dB) is the change in level that is just detectible. It is 1 dB out of the 120+dB our brain can perceive! That's 1,000,000 to 1 in sound *pressure*, which squares to 1,000,000,000,000 to 1 in sound *power*. Thank goodness our perception is logarithmic! So an overall system frequency response that deviates in precision by one decibel up or down (±1dB, 12% up to 11% down – again, its logarithmic) is quite acceptable to the ear\brain. Smaller deviations are designed for each link in the audio chain not to total overall a ±1dB alteration in tone color (timbre), the holy grail of audio.

---

[41] For humorous takes on our technological world, I refer you to engineer & standup-comedian Don McMillan.

In the vacuum tube era from 1920, 10% distortion was the benchmark of audio quality for engineers and manufacturers. It sounded over-bright but just shy of harsh "sibilance." In the 1950s with the transistor and the popularity of *hi-fidelity,* 1% became the new precision. Reducing 10% to 1% is a 20dB improvement. As a 10dB change is perceived as either doubling or halving volume, distortion reduced by twice 10dB seemed ¼ as loud, again in logarithmic perception. In the 1970s, integrated circuit "chips" decreased non-linearities another 10dB to 0.1% and lower (although speakers remain 1~10%). Compared to 1940s fidelity with 10% distortion, 0.1% is attenuated 20+20 =40dB, heard as 1/16 as loud in bogus artifacts. At this level, even playing music at its original loudness, the coloration (distortion) may be masked by environmental noise. Hobbyists may seek this goal, it they know how to (can measure it). If *you* didn't, you do now.

As with many physical phenomena, the dB is a measure not of perfection, but of precision that is *good enough.* Similar log-based measures are used for video, image exposures, musical intervals, earthquakes... The dB tells what you will likely hear: the full range of reproducible frequencies in *Hertz* (Hz, named for Heinrich, in cycles per second) within ±_dB deviation from "flat" tells preservation of tone color. The deviation in watts of amplifier power or speaker SPL spanning the full range in Hz at _% harmonic and comparable _% intermodulation distortion with all channels driven (no listener fatigue, or speaker-destroying clipping [42]). The signal-to-noise (SNR) ratio in dB down to hum & hiss. The dynamic range (DR) in dB hearable well below noise. Distortion <0.01% is 80dB below desired sound, and well below noise at standard 85SPL listening levels. As said, over-building the audio chain is unnecessary; under-building risks enjoyment.

Any component that is offered without meaningful performance data in dB (or its equivalent in %) urges looking elsewhere. Omitting, obfuscating, or falsifying the measured specifications will likely mean the manufacturer is hiding the truth. And there will be no quantitative reference for you to claim a lemon under warranty*!* For example, a meaningful frequency response is given as "__~__kHz ±__dB." But frequency "range" *without dB* is meaningless, and usually implies –10dB or worse. Buyer beware*!*

As with other quality measures, the mere *appearance* of "__dB" in specifications connotes a quality product. dB in published specs suggest the manufacturer is being truthful, as others could measure and possibly refute their claims. For listeners, the dB suggests the difference between a distortion-induced *ear-fatigue* v. lovely transparent sound. The effect is applicable to listening via PA system at a concert, to a movie in a cinema, or to records at home. The following may be just what you need to know.

---

[42] If symmetrical, clipping (Gibb's Effect) produces only odd harmonics beyond 100 times the fundamental. E.g. a clipped 5kHz tone's lowest odd harmonic is audible at 15kHz, but one at 101x is 505kHz, at the door of AM radio, that despite being proportional to 1/f (–40dB) triggers trouble. A simultaneous tone at 3kHz adds 9kHz at –10dB, more 15kHz at –14dB, and 21kHz at the door of CD sampling's Nyquist filter, etc., adding to RFI & IMD.

Measuring sound (science alert)

The decibel (dB) & human listening

Like the slide-rule (mechanical analog computer) engineers once used to add or subtract the lengths of two wood sticks, the dB allows multiplying & dividing by adding & subtracting. (You could discern the slide-rule's answer if you'd already guessed it.) The dB is useful because human perception, e.g. the senses of hearing & vision, is by nature exponential (logarithmic). Especially for audio, this section makes the dB understandable, and handy.

The *Bel* honors Alexander Graham Bell, whose telephone birthed the field of audio. Not just a physical unit of measure, the Bel is analogous to the logarithmic human perception of hearing. The Bel is any *change* in acoustic power, or its electrical analog, that signals in our brain a halving or doubling of perceived loudness. And the deciBel (dB), a tenth of a Bel, happens to jibe with a psychologist's "just noticeable difference" (JND) between two sensory levels. dB are useful in hearing and audio as a level difference, a level change, or a level relative to another as a fixed standard. dB-measured performance specifies how well a technological device works, whether it needs fixing, or where there's room to improve it.

Vive la différence...

A number of dB denotes a *difference*. Two levels are compared, at two points in an electrical circuit, or at two frequencies, for any device under test (DUT). +__dB describes amplification\gain, −__dB is a loss\attenuation. Using the exact formula below for voltage changes, ±20dB (+20 or −20dB) is a factor of 10 increase (gain) or decrease (loss). For example an RIAA phonograph pre-amplifier might need a gain (at 1kHz) of 100, or +40dB.

One of the compared levels can be a fixed reference, then dB adds letter(s): dBw, a level above or below 1 watt; another is dBµv a measure in micro-volts. Sound pressure level (dB$_{SPL}$) refers to a change in air pressure that at its minimum of 0 dB$_{SPL}$ is the "threshold of hearing." [43] In audio, the electrical analog of sound, the level standard of the Institute of High Fidelity manufacturers (IHF) is −10dBv, or 10dB below 1 volt, =0.316v or 316mv (milli-volts). In the 1920's, telephone systems used dBm, the power dissipated by a load of 600Ω, which for its reference 1mw produced 0.775v. Today in professional audio that same voltage is termed dBu (sometimes dBs), so 0dBu refers to 0.775v rms. For a cartridge that outputs −50dBv, the preamp gain needed is 40dB (100x) to reach −10dBv "line level."

The dB for measuring voltage or power changes, simplified...

Less than 1dB goes unnoticed because it's less than a JND. Noticeable is + or − 2dB. ±3dB (71% or 141%, the f_3 of filters) is obvious but not ruinous. A frequency response that is supposed to be ±0dB (flat) but is ±4dB or more changes timbral color. In stereo, 15dB difference between channels pulls a sound fully to one side [Theile]. 2dB shifts a soloist quickly off-center, creating a "hole-in-the-middle" whenever a sound moves (is panned).

To *measure* dB levels with respect to either 0dBv (1v) or 0dBu (0.775v) standard, we use an AC voltmeter (ACVM) with a flat frequency response and with scales for dBu or dBv (most meters have both). Then to *calculate* a __dB change in sound pressure (SPL) or voltage (electrical pressure), divide the SPL readings or two "pressures," V_1 & V_2, find the logarithm in base 10, and multiply by 20. [44] A change of 0dB is unity (x1, no change).

---

[43] The threshold of hearing is 20µPa, 0.000020 Pascals, N/m², or any change in air pressure of 0.0001450psi.

[44] dB=20x log$_{10}$(V1/V2). Using a smartphone's calculator app, the keypresses are: V_1, ÷, V_2, =, log$_{10}$, x, 20, =. Because from Ohm's law a √2 change in voltage doubles or halves power, two wattages use 10 x log10(P1/P2).

Many dB calculations can be done in your head by simply adding dB to represent multiplying signal levels for a gain. Subtracting dB to represent dividing signals for a loss. For centuries this is what a slide-rule did. You already know +20dB is a 10-fold gain (because we multiplied by 20 the $\log_{10}$ of V1/V2). Reducing to 1/10 is a loss of −20dB. And that amplifying 100x is 20dB+20dB =40dB. Adding 40+20dB multiplies by 1,000.

Double the signal voltage is close to +6dB; half is −6dB. So 4x is 12dB; ¼ is −12dB. The two audio references, 1v and 0.775v, differ by about −2dB. Then the difference between IHF standard −10dBv and professional standard 4dBu is about 14−2dB. So consumer audio voltage levels are about −12dB lower than professional voltage levels. [45]

To figure in your head a 5x gain, it's half of 10x. In dB that is 20−6, or 14dB. And a loss to 1/5 is −14dB. Got it? How many dB is 20x? What amplification is 54dB? [46]

Just two more practice exercises for calculating dB in your head: +8dB is 14−6dB, half of 5x, or 2.5x; −8dB is the reciprocal, 40%. ±2dB is 8−6dB, or half again, so about 1.25x, and −2dB is about 80%, (precisely 79%). ±¼dB, only about a ±3% change, is inaudible.

Now you can mentally figure close enough a ±dB change or standard level, or use the scale below, estimating by the lower voltage ratios. Power changes are half as many dB. But pressure or power, the effect in our brains of a certain number dB change is the same.

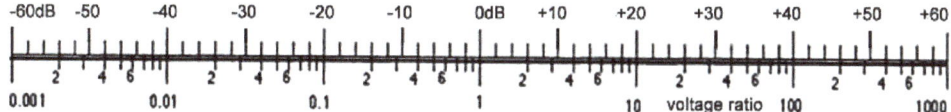

What dB magnitudes & variations mean to you?

Applied to the phonograph, "flat" is determined by C-load, in dB on **p74**. The RIAA characteristic is on **p86**, and **below** ±¼dB for the project preamp, "rolled-off" above & below the audible range to avoid noise. Flattest are amplifier stages, often ±$^1/_{10}$dB.

Once the numbers become familiar, measurements reveal how a component or transducer will sound – to what extent it will modify the timbre (tone color) of sounds. Note that bass sounds will be perceived as timbrally weak if played softer than original, in dB of loudness, or booming if louder (highs are exaggerated less) [Fletcher-Munson, **p101**].

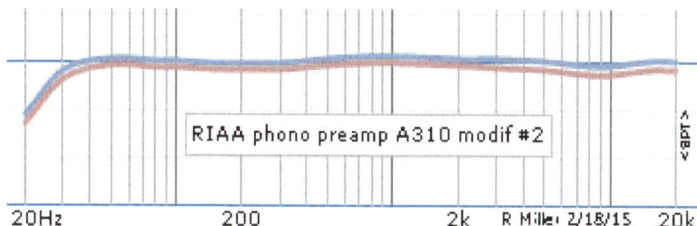

Frequency responses are less than ±¼dB for the preamp project (L & R separated $^1/_{10}$dB for clarity).

---

[45] Analog reference levels and "line-up" tones reserve "headroom" (HR) for dynamic peaks up to 20dB higher than average level at which *overload* occurs. "Vinyl" and magnetic tape HR is ~12dB. For digital recordings, reference is set ~16dB below "full scale" (FS) at which *clipping* occurs, allowing HR of 16dB. Movies in cinemas have lifelike dynamics with HR is 18~20dB above the 85dB$_{SPL}$ reference [SMPTE\EBU]; dialogue normalization is −31dB below FS of 105 dB$_{SPL}$ emulating normal conversation of ~74dB$_{SPL}$. Least natural, over-compressed pop radio and CDs squash dynamic headroom to as little as 3dB, and intentionally clip 4% of peak samples.

[46] 20x is 10 x 2, or 20+6=26dB. 54dB is 6dB less than 60dB, which is half of 1,000, so it is 500x.

The frequency response **above** shows in dB the phono preamp project later in this book, signaling fine quality despite its low cost for a few readily available parts. Costing 50x as much (+34dB$!), **below** describes the performance of another transducer, one of earth's finest microphones, only 3/4in (19mm) in diameter and a couple inches (50mm) long. Even with a critical mechanical link, its frequency response is remarkably linear (flat). [47] The tilt below 200Hz is intentional to prevent strong bass *proximity effect* when used close – it is flat at 12in (30cm), or requires a LF boost at farther distance, or cut if used closer.

Particularly telling are the rings of the *polar response* diagram, separated 5dB, that reveal that its frequency response is quite neutral (concentric) in tone color for sounds not just in front, but from every direction, actually in 3D as though the diagram is spun about its axis. This means it has very real-sounding pickup for concert hall music and natural sounds in reverberant spaces, and superior "reach" cf. shotgun units for film/video dialogue pickup.

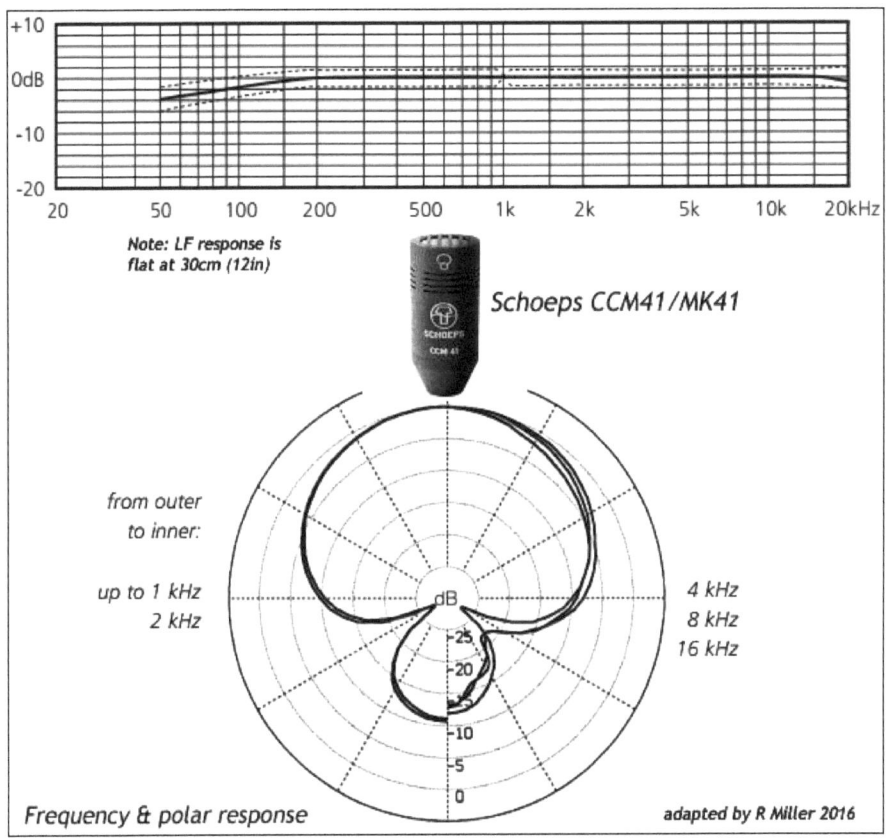

One of a family of microphones you've heard many times in music recordings and movie dialogue, the sound of which is revealed in dB-based graphs. Its frequency response is within the dashed ±1½dB manufacturing variance. This "hyper-cardioid" is filtered <200Hz for *proximity effect* boost when used close; at 12in (30cm) it is ruler flat 50~20kHz, down to 20Hz for omni-directional versions. The polar chart (a sphere when imagined rotated about its axis) shows it is flat (uncolored) for sounds from any direction in 3D. At nearly any angle at all but the highest frequencies, response varies ±1dB, which most cannot discern. A marvel of engineering, here given *imprimatur* by the dB. Courtesy Schoeps Gmbh. [The author receives no compensation for any content.]

---

[47] A "transducer" has the tough job of converting one form of energy to another linearly (undistorted): for a phono pickup, mechanical vibrations to electric signals, sound to an electric signal for a microphone, sonar sensor, etc.

Sound Perception and better sound from your digital

Even before you lay eyes on any performers, you recognize the music is *live*. Yet after a century and a half, audio reproduction still falls short of *verisimilitude* (true-like). Auditory reality demands lowest distortion, and *immersion* in the *3D sphere of hearing*.[48] Not the single channel of *1D* monophonic, nor the *1½D* of 2-channel stereophonic speakers pie-slice-shaped horizontally in front. Not even *enveloped* in the horizontal *2D* circle of 5~7.1 surround, nor *semi-immersed* in the hemisphere of *2½D* Dolby Atmos.[49] Then, one adds his\her own sonic opinion (taste) that forms one's "real reality." So an individual reader's *current* (but changeable) perception also plays a role in Better Sound for your Phonograph.

What happens in our brains listening to sounds is the science of *psychoacoustics*, the psychology of *live hearing*. Introduce an interloper, audio recording & reproduction, begs the question *why the variety of individual subjective perceptions?* One's perception may be altered by new knowledge that might change *taste*. Except with repetition, it has become engrained opinion, belief, and closed-mindedness that only seeks agreement, known as *bias confirmation*. Initially, perception is *nature*; in the steady state, it is *nurture*, the result of conditioning and *habituation*. And in time without an open mind, harder to change. The oft-chanted claim is "It's personal choice." Though we are also free to learn and change it.

At any given time, are our subjective expectations met? Exceeded? Or does the quality fall short? Individuals refer new experiences to the sounds in their past, especially during formative years. Children of the hi-fi era of the 1950s, 60s, & 70s, my spouse and I enjoy mostly classical & American Songbook jazz albums. And we became semi-pro musicians (voice & piano) of acoustic music. If today I were to hear that earlier era's audio quality, I'd likely think it awful. But instead I remember that *then it sounded good*. Somehow as good as the far better quality playback I have now*!* In their time, both met expectations.

I know both eras' qualities cannot compare; that each improvement since has recalibrated my expectations. My perception updates whenever I first experience an evolutionary leap in quality. With 60+yr "ear training," the slightest change, up or down, is accompanied by a wow-factor. Like Edison's 1915 Diamond Disk Challenge when attendees were wowed when they could not discern live musicians on stage from a recording of them.

Just before this 2nd edition went to press, my spouse and I attended a family reunion. I settled on the porch with the next generation of mostly guys, 40-somethings speaking in tongues about video games and such. Then suddenly they were talking about "Frank." Sinatra that is. They raved about "trombones." And the "bass." They'd entered my wheelhouse, tho I kept silent, fascinated by where the conversation would lead. Without a shred of technical analysis as in this book, they'd discovered the vast repertoire of the American Songbook. Acoustic music of the Big Band era. The love songs of George Gershwin. The sexy poetry of Cole Porter. The energetic arrangements of Duke Ellington, Count Basie, and Nelson Riddle. The interpretations of Ella Fitzgerald, Mel Torme, Peggy Lee, and so many others. The many covers of these masterpieces by contemporary artists. Well recorded acoustic jazz music, from master tapes, but also restored from 1950~70's vinyl by conservator services like the author's – at a level of replay quality available to you.

---

[48] Hearing gathers sonic arrivals from a full-sphere, including reverberation that completes tone color (timbre).

[49] Even in demonstrations of the author's 10\14\26.2 speaker full-sphere HSD-3D – filmmaker.com/papers.htm.

Unlike the readers of this book, they were consumers of the music, but not (yet?) aware of how best to play it. They'd also discovered dynamic audio. Listening to this music over satellite radio in their cars, it's lightly compressed to overcome road noise, if not previously smashed to the ceiling in the mix. And over Bluetooth and smart speakers with tiny drivers physical capable only of "resultant" bass from the fundamental's unique overtone structure. Avoiding the lowest octaves where acoustic treatment is most needed. These are not audio enthusiasts, but the majority of ordinary music lovers, where enthusiasm for audio ought to begin. Will they next discover classical, I wondered? Where the lowest "bass" requires that acoustic treatment, and low-distortion subwoofers capable of reaching the 30Hz of C-extension double bass viols, the contra bassoon, tuba, piano, organ, and 40in bass drum, reinforced all the way down by the reverberation of a large concert hall. This is the highest calling for any sound recording, analog or digital. In libraries and private collections just waiting for proper reproduction, including vinyl that has no trouble rendering 30Hz cleanly.

A peer reviewer asks why playing 78s today he prefers the sound of an acoustic Victrola, perhaps conjuring what his (grand)parents enjoyed? Might the younger generation at our reunion somehow be conjuring the 1950s~70s sound recorded long before they were born?

When I hear "How come records sound better than digital?" I try to understand: "Where does this come from?" Have they not heard good digital? The claim refers to the failure of artificial in-your-face digital releases cf. the "warmer" sound of an original LP. When the initial ear candy turns to a sugar low. Soon the market might come to suspect LP reissues, often mastered from an over-processed CD! Less processed vinyl ultimately sounds better.

No matter the genre, to what degree does enjoyment of music depend on audio quality? Thus far we've considered quality of replay. Because readers of this book can control that. We've assumed good recording quality – a big assumption. [50] A whole other book might delve further into what precedes replay in cut grooves, on master tapes, or in digital files.

A digital recording should sound superior unless it has been poorly made, which is often the case. Compared to their excellent original releases on vinyl, Beatles' Abbey Road is to me as good on CD; Fleetwood Mac's Rumours isn't. It, Blood, Sweat & Tears, and Giles Martin's Sergeant Peppers are transcendent on vinyl; many recent releases, not so much. It would sound better if digital recordings were as well-mastered as vinyl in its Golden Age. As good as most new releases of acoustic performances, for which we have a live reference.

Among other reasons why digitally recorded popular music may sound poor, perhaps worst is over-use of processing. As the music business shifted emphasis from *the song* to *the sound* [Horning 2013], compression was used so individual instruments would not be drowned out amid the layers overcrowding popular music's wall of sound. Most artificially affecting acoustic instruments, compressing classical & jazz is usually subtle, if used at all.

Popular music in digital as distributed is often over-processed for lowest quality replay systems, but that on high quality systems sounds bad. Vinyl in hi-fi's Golden Age was mastered by engineers who learned how to work within vinyl's limits. With the opposite goal of sounding natural by avoiding the distortion mechanisms inherent in grooved media. And the limitations of vintage electronics. We've learned earlier in this book about phonograph pinch effect, tracking angle distortion, "poid"-like distortion, monauralizing below ~150Hz, skating, plus rumble and surface noise. This book can help readers reduce these artifacts to optimize technical quality, which in turn enhances enjoyment of the music.

---

[50] In the author's collection, most acoustic music (classical, jazz, etc.) on both vinyl and CD are well-made.

We all listen to what we like. But while high-definition video displays seek more natural "retinal" reproduction, of chiaroscuro bright & dark, must sound be robbed of lifelike dynamics? Picture display dynamic limits are not "scaled" in the same way as audio reproduction – there is less ratcheting up of costs as with power amplifiers and speaker drivers. Instead, audio content is commonly processed for lowest common denominator gear. But if it sounds horrible on better gear, why invest in it? And why, when naturally dynamically mastered releases could easily be compressed as needed in listeners' receivers?

Digital sounds best if its processing is not overdone

For vinyl up to 1982, processing was not a bad thing unless it was badly done. The "Volume Wars" on radio escalated, and was too easy to do digitally after introduction of the compact disc. With 110+dB of dynamic range, including audibility well below noise, the CD needed little if any level compression. It got worse after "normalizing" it to the ceiling, then increasing the level further until digital audio samples are clipped, never mind that flat-topping a wave generates the distortion percentages of 70+ years ago. Competing labels followed. The waveforms **below** illustrate this manipulation. A nicely dynamic digital recording at the top is also typical of it on vinyl. In the middle for digital release, it has been normalized, then increased further until samples are clipped, shown in the closeup.

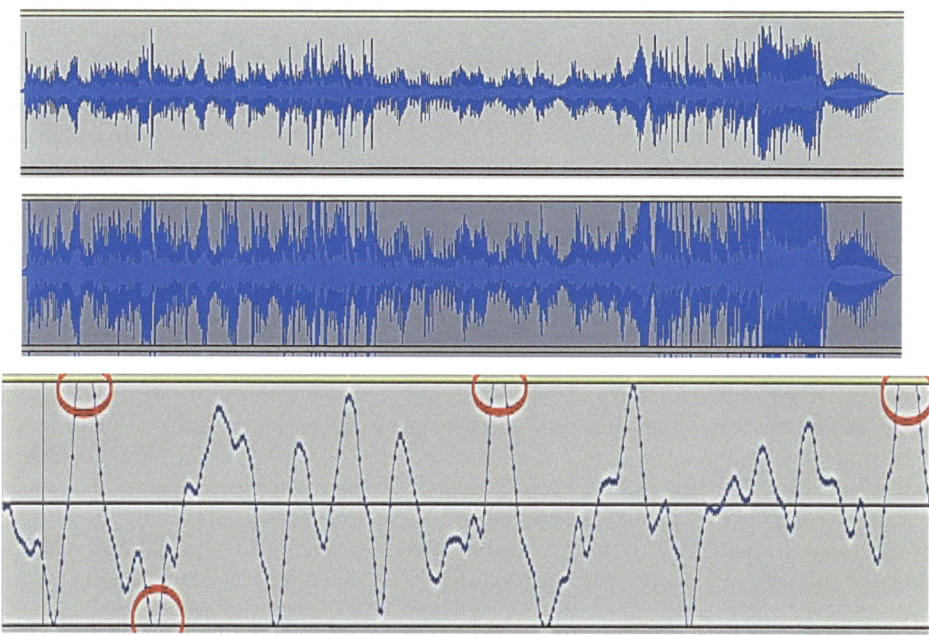

Top: original uncompressed (naturally dynamic) recording, also typical of vinyl. ***Middle:*** same recording after smashing to the ceiling using excessive level compression – little is allowed to be naturally softer or louder. ***Bottom:*** close-up of nasty-sounding "clipping" of peaks & valleys.

Sound can be played louder without destroying dynamic "punch" simply by turning up the volume. But label deciders know consumers don't do that skipping through songs to buy. Labels know that an instantly louder song catches attention. *Commercialized sound has transitioned from musicians' art and engineers' craft to marketing strategy, overtopping prior limits. Where long-term value for the customer takes a back seat to the instant the sale is consummated.* Label executives have decided for consumers that nothing is to be naturally softer, nor naturally louder. The close-up **above** shows clipping causing spits of hash. Unaware, most consumers fall for the con, although many become dissatisfied later.

Other audio processing adds hours and days mixing & mastering to create artifice, define a novel aesthetic. Copy-&-paste loops lack human feel. Click-tracks mean expressive *rubato* is nowhere to be found. Where the music created is only heard over a loudspeaker. Vocals deprived of bending and vibrato by AutoTune, though useful when out of necessity. But for those who recall the fresh ingredients in the recipe for music of vinyl's heyday that simulated live performance, today's studio product sounds canned – as disposable as 78's steel needles. Today, live concerts emulate performers' studio recordings, not the other way around, as in the past. Concertgoers have come to expect to pay high-ticket prices to hear less than CD quality over a PA system, having become conditioned (habituated) to listening only to over-loud genre designed to overpower escalating street noise or dance club conversation (yelling). Many acquire a taste for the over-stimulation. Subtlety has given way to incessant overcaffeinated energy. Art super-sized. Art for commerce.

The 140+yr history of vinyl is largely intact, because a stylus would jump the groove as transducers cannot negotiate the right-angle corners of clipped waveforms that hardly inconvenience bytes. Same for inertial loudspeakers & aging inner ears, as they too are analog transducers. Add that good content almost always trumps the technology of its day, as proven by listening to a wonderful rendition on an ancient 78 or ET that survives. Undeniable are the inconveniences of old-fashioned analog record-playing compared to the ease playing new-technology digital. And now merely asking your smart speaker to play a song by title.[51] But many find they like the ritual of delicately placing a gemstone within a half-mile spiral. Setting aside multi-tasking; taking time for pleasurable *focused* listening.

For any well-recorded media, quality issues (distortion) overwhelmingly occur in replay. Not baked-in in the vinyl disk you are about to risk enjoying or damaging. Unlike digital, users can optimize disk reproduction by applying effort and scientific knowledge – better sound is up to you*!* This 2nd edition's title is "Better Sound *from your* Phonograph." Then consider 140+ years treasure of recorded history (70+ from the Hi-Fi era >1950) – gems preserved both in vinyl & pre-hi-fi shellac. Many reissue CDs, downloads, & streaming originate by "ingesting" from records using techniques on par with what you could have. However, well re-recorded digital can't be beat. *And it can serve well mastering new vinyl.*

Regardless of your chosen technology or taste in music genre, the essence of this book is to improve reproduction accuracy. For multi-mic'd, multi-tracked, multi-processed popular music, that means "extending your ears" back to the mixing desk to come close as possible to the sound the producers and musicians intended in their mix. For minimalist recording of acoustic music, it means extending your ears back further to the live concert. Because no recording is perfect, especially one captured live, more accurate reproduction will mean revealing it warts and all. Even with a few imperfections and crowd noises, the artistic energy of musicians in live performance is difficult to achieve multi-tracked in a studio.

If you make either DIY project, either the accurate RIAA preamp or the low-distortion transcription tonearm, then you will likely hearing better sound that does not reflect their low out-of-pocket costs. And if you already own the latest and greatest, the information herein is as valuable toward optimizing your system's return on a likely greater investment. Raised on the best reproduction his dad had available at the time, my 40-something son's taste drifted to level-compressed content over data-compressed streaming (the horror*!*). This until he was reintroduced to vinyl's qualities after making the tonearm in this book. At first, he expressed dislike for the cleaner sound from a vintage disk. His mother voiced her incredulity; and after listening to only a few more minutes, he had an "aha*!*" moment.

---

[51] E.g. "Alexa, play Bethlehem Progressive Ensemble" to hear from BBC-2 one of our jazz recordings from 1965*!*

The bugaboos of grooved media, often under enthusiasts' radar, are level-dependent 2nd harmonic distortion due mainly to *pinch effect,* or all-order harmonics from *poid* sawtooth artifacts. These are caused by stylus shape, as investigated from **p11**. With magnetic tape, non-linear *hysteresis* causes odd-order harmonic artifacts – peaks have 3%. When used as a source for cutting disks, the master and prior tracking tapes add odd-order distortion to disk playback's mostly even-order. Friction causes skating on vinyl; *scrape flutter* on tape. Loud passages *echo* by print-through on tape; by "horns" raised aside grooves by heated cutters. While some of these can be optimized, digital reproduction suffers none of these.

Digital audio has advantages in most ways, evident in well-recorded jazz, choral, folk, and classical that are perceptually demanding. Acoustic music releases generally are not compressed for *lo-fi* gear with limited dynamic capability and often played in high noise environments. They preserve natural peaks in level that exceed the average – *headroom* (HR) reserved for natural dynamics: ~12dB for *peaks* on tape\vinyl, 16dB for CDs, 18 or 20dB for cinema's lifelike sound. Over time, HR was hijacked to escalate sounds levels, often leaving for dynamics only 3dB (measured)! Those with better equipment in quieter environs reject this worse sound, so buy less music. With 96+dB bits, digital doesn't need it.[52] In the 50s~70s, 45rpm engineers tacitly agreed to master singles at the same volume, so juke boxes required one setting by the barkeep, per C. D. Krepps [interview **p141**]. [53]

The other "compression" – digital data reduction – tosses up to 90% of audio deemed redundant. Perceivable by many, for most ear-bud listeners it's an acceptable trade-off for cramming more services down digital pipelines. At low background music levels, artifacts can be drowned out by *environmental noise.* But at cinema or audio enthusiast volumes in quiet theaters and listening rooms, distortion artifacts of all but high bitrate audio cross the noise threshold to become audible. Irritable for some; for many unconsciously fatiguing.

"HD Radio" (in band on channel, IBOC) embeds compressed (data-reduced) digital audio. FM multipath distortion and dropouts while driving are "fixed" by *latency* of a few seconds in a memory *buffer,* also used by the internet, satellite, WiFi, and Bluetooth. It's the cause of several seconds of silence when the buffer runs out. Satellite radio fails in rain, when trees have leaves, and in tunnels without repeaters. Man-made radio frequency (RF) pollution is rising, so all RF-based communications is chasing its tail. Lifelike sound for discerning consumers is still a dream. But good content can overcome a lot of distraction.

LF & HF "f_3," cutoff frequencies, noise, & distortion in perspective

The most critical audio chain links are phono pickups and loudspeakers, both transducers. Distortion generated by one link early in the audio chain is re-distorted in latter devices. So more critical is the earlier stylus tracing – why we dealt with it first and extensively here. For purposes of this discussion, we'll assume both the groove as mastered and the listening electronics are far more linear. Also recall that sound (and thus audio) are by and large an instantaneous mix of harmonically related sine waves [Fourier], plus a smattering of noise and intermodulation sum & difference "heterodyne" tones from cymbals, vibraharp, piano, etc. We've shown the harmonic artifacts added to the original signal by stylus mis-tracing, manifesting as false brightness up to unacceptable "sibilance," along with added heterodyne artifacts manifesting as a "burr" or gong effect, each measurable at up to 10% distortion for a 45\45 stereo groove. In the end, the speaker generates similar percentages. However new

---

[52] More than signal-to-noise ratio (SNR, or SINAD including distortion), dynamic range (DR) extends to audible sounds 15~20dB below sampling noise. A 16bit CD with ~93dB SNR can have an effective DR of 108~113dB.

[53] Yet Krepps' breakout was Manfred Mann cover of *Do Wah Diddy Diddy,* "the loudest record ever recorded."

added harmonics and heterodynes affect the prior harmonics and heterodynes, resulting in artifacts upon artifacts unrelated to the original signal. Now the audio is tainted by unreal bogus sounds, and verisimilitude is lost.[54] Now evident is the importance of catching non-linearities at their earliest occurrence, at the stylus. Even the just as copious speaker non-linearities (distortions) at the end of the audio chain are more benign. High levels of added HF – ultrasonic out-of-band harmonics and IM sum tones – can be reduced with a low-pass filter (LPF) with a cutoff $f_3 \geq 16kHz$ built into an amplifier, but IM difference tones remain.

Changing one form of energy to another, microphones, loudspeakers and phono pickups have the most difficult jobs in audio, reflected in their costs to make and buy them. Their in-band frequency response measures their deviation in tone color (timbre). Variation in dB can tell how they can be expected to sound, even before you try them. If performance specifications are not given, including references, assume the unit has been designed to a *price-point,* not a *quality-point.* With speakers, an advertised "frequency range" implies – 10dB points at their half loudness limits of lowest and highest frequencies, or worse. Monitor-grade speakers specify where response falls –3dB ("f_3's") between which some vary only ±1dB, and like the microphone on **p94**, reveal graphically their polar dispersion.[55] An f_3 of –3dB sees voltage reduced to 70.7% (50% power), an audible but not as critical a loss as –10dB that sounds like half loudness. In audio, these "f_3" points define the *filtering action* at the limits of a device's performance, often intentionally applied to avoid trouble.

Frequency response dips in are less noticed than equal bumps: –2dB is less objectionable than +2dB humps that stand out. Dips\bumps of a series of components accumulate in dB. So best practice is to engineer out bumps, tricky for transducers. For amplifiers facing few demands (unlike preamps & power amps) to set their f_3 extremes below\above audibility.[56] This is why proper phonograph replay attends to cartridge C-load and accurate preamp EQ.

Noise is measured in dB below a reference level, the *signal-to-noise ratio,* or SNR. Uncorrelated noises do not add linearly as do correlated signals, but add statistically from studio AC, mic preamps, magnetic tape, replay preamps, and the listening environment. 60dB is a typical end-to-end weighted SNR of attentive home listening.[57] E.g. the range from a loud 90dB SPL to room noise 30dB above the threshold of hearing (NC30). Or a typical vinyl release with 60dB SNR between distorting overload and surface or preamp noise. Less audible LF noise is weighted lower to agree with the decreasing sensitivity in Fletcher-Munson re its perceived equal at mid-frequencies, where the ear is most sensitive. Within human hearing's dynamic range (DR) of ~120dB, we can hear intelligence 15 to 20dB below noise! So 110+dB of DR is possible on a CD; 70+dB for vinyl. Softer pickup of ambience by the microphone on **p94** 15~20dB below noise preserves dynamic realism.

Distortion is also audible below noise. And becomes objectionable whenever it exceeds noise. An overall distortion of 10% (–20dB) from a vinyl disk, AM radio, or low bit-rate streaming *might* fall below audibility amidst car or street noise. Or be OK as background. But in a quiet cinema, 1% distortion (–40dB) is not masked by noise, unnaturally coloring

---

[54] (Sub-)woofer THD creates bogus "resultant" low bass cf. **p102**; mid-range ("squawker") artifacts in the ear's most sensitive range affect solo voices; tweeters' UHF harmonics are inaudible, but not difference heterodynes. Lower artifact %'s at levels around threshold unnaturally come & go, varying with SPL, higher noise masking, or competing sounds in the same *critical band* of hearing (a "spectrometer" with 24 cochlear *Barks* spaced 0.337f).

[55] The family of microphones represented on **p94** has a distortion spec of 0.5% at highest SPL level of 131dB.

[56] Intermediate flat amplifier stages typically cut off LF at 1.59Hz, HF at 159kHz, and vary <0.1dB 15.9~15.9kHz.

[57] Twice overall SNR of audio reproduction in practice, 120dB is conventionally the full range of hearing – "more than the difference between a mosquito in the same room and a jackhammer one foot away." [Monty 2012]

sound that distracts from movie goers' *suspension of disbelief.* Relaxed ~10dB for home theater certification, THX for cinemas requires controlled acoustics, noise below NC30, and an undistorted 105dB$_{SPL}$ for each channel. Hobbyists may desire comparable quality, but rarely achieve it up to 95dB$_{SPL}$, stopping where distortion says "it hurts; it's too loud!"

At the perceptual cost of reduced realism, dynamically compressed recordings reduce the cost of consumer electronics. E.g. mixes with only 10dB headroom require 1/10 the power capability of consumers' power supplies, amplifiers, and speakers. Say that is 10w max. 20dB of HR that preserves the headroom would require all audio components handle 10dB more signal, for undistorted peaks of 100w for which power chips & speakers cost a lot more. (105SPL in cinemas even with higher efficiency speakers need 500+w per channel.)

Unknown is whether labels will ever reinstate naturally dynamic releases. We believe, given the opportunity, that many consumers would gladly *re-habituate* to natural dynamics. But if one rarely attends live unamplified concerts for reference, then s\he likely believes nothing is amiss s\he'd pay to change. This book's exploring flat frequency response and lowering noise and distortion anticipates the market will awaken to the greater enjoyment of artifact-free sound and correct tone color (timbre), the holy grail of *high fidelity*.

How loudness perception differing from the original distorts tone color (timbre)

Before charging off on a trial-and-error quest for your *preferred sound,* you might begin with "neutral" reproduction that conveys as accurately as possible the intentions of a disks' producers. Especially for acoustic instruments (piano, drums, acoustic guitar, human voice) for which we have a remembered reference for realistic timbre (tone color) that has, since the 1940s, been the holy grail of *high fidelity* (hi-fi) sound. But many factors can derail it. Least suspect nowadays are active solid-state devices that have become very good even if inexpensive. Most challenged are transducers: microphones, loudspeakers, phono pickups – allocate at least half your hi-fi budget to these! Then add more for *acoustic treatment.*

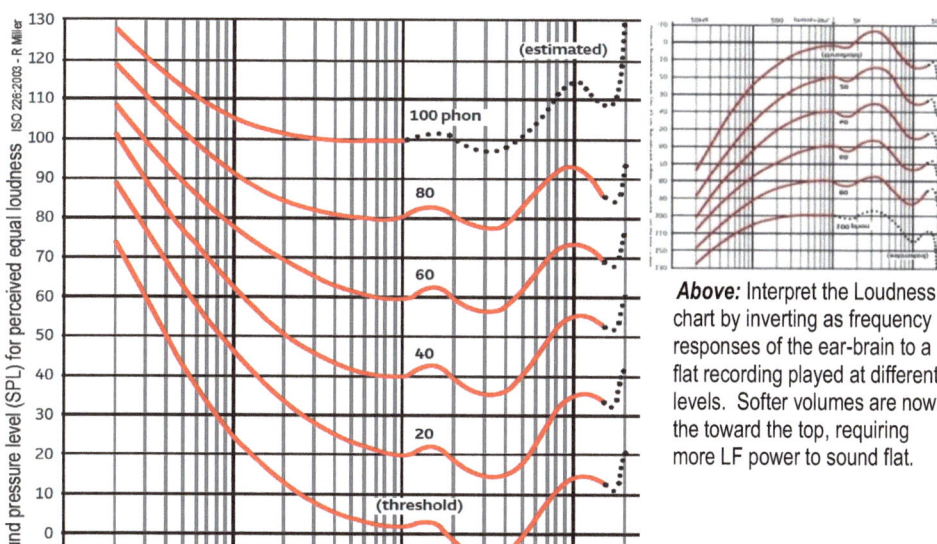

Above: Interpret the Loudness chart by inverting as frequency responses of the ear-brain to a flat recording played at different levels. Softer volumes are now the toward the top, requiring more LF power to sound flat.

Above: Int'l Standards Organization 226:2003 updates Fletcher & Munson's 1932 work at Bell Labs, how hearing filters and converges sounds below 700Hz. Absent the needs of migratory birds; sparing humans hearing barometric weather changes! At softer than original SPL, bass is perceived as weak, requiring more LF power to sound "right." But subwoofer distortion can be louder than barely audible fundamentals!

There is another commonly overlooked issue affecting tone color – *listening level*. The movie industry takes pains to standardize the sound pressure level (SPL) in theaters and mixing studios. Because if replay volume differs from that monitoring the original mix, it can dramatically alter timbre (tone color). First published in 1933 by Fletcher & Munson, the Equal Loudness contours are illustrated in the **previous** graphs. Related but not the same as audio level, *loudness* measured in *phons* is the *intensity of the sensation of sound*.

The **red curves** (as perceived in *phons*) are statistical measures of the ear's sensitivity v. frequency at that loudness (in dB SPL). Really a continuum, phons are scaled the same as physical SPL, drawn equal in dB at 1kHz. But when replayed at a different SPL than recorded, the original's frequency response is not heard as intended. Note how the curves converge at lower frequencies. For example, orchestral music originally performed, recorded, and mixed up to the 100phon curve but played as background music 40dB softer is heard following the 60phon curve. For low bass instruments (double bass viols, piano, tuba, contra bassoon, bass drum) reaching 30Hz to sound "real" requires 20dB higher SPL or bass will sound 20dB weak*!* The phonograph is quite capable of flat response to 30Hz.

20dB weak is perceived as ¼ in loudness. Conversely, played louder sounds booming. Altering ratio of a sound's LF fundamental to its HF harmonics distorts its tone color. In the hi-fi era, a *Loudness* control was included in amplifiers\receivers to preserve tone color playing softer than the original mix, after calibrating to a reference SPL using the Volume control. To boot, LF distortion harmonics may be audible *even if a bass fundamental is too weak to be*. The fundamental may be perceived only as a "resultant" inferred from its harmonic structure. Many audiophiles may never have heard low bass fundamentals*!* This science is revealed in the short paper "Subwoofer Camp" at www.filmaker.com/papers.htm.

Habituating to coloration (distortion), level compression, and no low bass

High fidelity is defined as faithful to the original sound, after optimizing mechanically and electronically the linear signal transfer (distortion) and noise within the audible range of 20~20,000 cycles per second (Hz, for Heinrich Hertz). Applying also to digital replay, choosing speakers with good dispersion, and treating interfering room reverberation.

Caused by any non-linear transfer of an analog signal, several forms of distortion reduce fidelity. Most talked about is percent total harmonic distortion (THD). Even-order (2nd, 4th, 6th...) harmonics are 2, 4, 6... multiples of the fundamental frequency they augment; odd-order ($3rd$, 5th, 7th...) are 3, 5, 7... multiples. Most sounds have natural harmonic structures – usually either odd-order only, or both even & odd. Altering its natural spectra changes a familiar sound's *timbre* (tone color), the holy grail of hi-fi. Tame HD for a similar percent reduction of intermodulation distortion (IMD), where a non-linear device causes any pair of tones (fundamental+harmonic, harmonic+harmonic) to form new sum & difference tones.

Most artifacts are playback issues that add artificial brightness that some audiophiles find compelling, although bogus. The strongest harmonic added is usually the 2nd, an octave, so more benign. A musical 12th, the 3rd, less so. IM's "burr-\gong-" like sounds are the worst.

Phonographic reproduction has many non-linear gotchas. Quality digital is plagued with fewer, until the crime of over-modulation is committed. Positive tops & negative bottoms of the waveform are "clipped," introducing odd-order harmonics at a far higher percentage than most forms in phono reproduction. For the duration of clipping and the ringing that follows, artifacts to more than 100 times the fundamental frequency add sounds like a spitting trumpet (already an odd-order harmonic producing instrument). Vinyl cutting and reproducing styli cannot negotiate clipping's right-hand turns without severe *ringing*. As background music, distortions may be softer than noise enough to be masked. But played on full-range systems at performance volume, these un-recorded imposters will be heard.

On **p97**, an egregious form of distortion is intentional: squashing loud & soft sounds together and to the ceiling is SOP for digital pop music and new vinyl releases and rereleases. Labels practice level compression: 1) to hear *layers* above environmental noise; 2) to create a high energy rush; 3) to drown out a lousy vocal; 4) to dominate in a competitive market. When a shopper compares musical selection after selection, the *louder* one instantly seems *better,* and the register goes ca-*ching*. The softer one is less level-compressed, thus more dynamic, but who while auditioning takes time to simply turn up the volume?

Over-compression reduces the ratio of loud to soft sounds from vinyl's 12dB headroom to as little as 3dB as measured in digital releases! Emptying the least of a digital's 16 bits to 12 usable bits, equivalent to vinyl. Depriving listeners of instruments natural highly dynamic nature: drums squashed to sound like flyswatters on a carboard box. (On one pop music session, the author recorded just that to prove that for most ears, including the A&R types in NYC we shopped it to, it was indistinguishable from real drums over-compressed.)

Some record executives order "4% of digital samples be clipped" to play compellingly on lo-fi systems, drowned in traffic noise, or as background music. Focused-listening by hi-fi enthusiasts may find the odd-order clipping artifacts more objectionable than the even-order HD distortion of vinyl. Distortion matters most (if not only) to hi-fi enthusiasts who crave not novel fuzz-toned audio, but "real" sound – verisimilitude. Compression should move from one-size-fits-all mastering to players, where a user sets it as needed (including none).

Compression allows budget players to be cheapened by using lower power amplifiers and speakers. For example, naturally dynamic motion picture sound preserves headroom peaks higher than average by 20dB [SMPTE standard, 18dB EBU]. This requires theater & home cinema systems have peak power capability 100 times average. Level-compressed to 10dB requires only 10 times the power capability for lo-fi consumer systems. Cinema amplifiers rated 600w v. 60w [58] for typical consumer audio systems; cinema woofers with maximum cone displacements of 10mm v. 3mm for boom-box players. Producers' practice of level compression goes hand-in-hand with audio manufacturers' downsizing-to-price.

As "euphonius" distortion coloration becomes some audiophiles' preferred brightness, humans habituate to it, and to level compression. But just as learned through exposure, undistorted reproduction and naturally dynamic recordings can be re-learned. Evaluate differences under controlled conditions, e.g. blind testing. Sighted tests delude participants to opinions instead of science, by *confirmation bias* that favors *looks* or *high-end* prices.

The sound quality of audio reproduction is adjudged in the listener's brain – *psychoacoustic perception.* "Audiophiles" often tinker by trial & error. And claim to "trust their ears." Or believe influencers in the echo chamber of advertiser-paid reviewers and online "opiners." These engrain confirmation biases. Affected by tiredness, mood, or mind-altering substances, trusting one's "ears" is unreliable. More fleeting than factual. Better is to be open-minded, to *be scientific.* My 60+yrs as an audio professional chronicle many stages of relearning, and I am not done. Human conditioning is essential for unconscious routine behaviors, but also why double-blind or ABX testing is required for meaningful "subjective tests." Why 30s is the limit for audio test clips – it takes only a bit longer for accommodation. Why a few minutes of well-reproduced audio can reboot one's opinion.[59] It is easy to add distortion, if coloration is what a user desires. Harder is to eliminate it.

---

[58] Amplifier power P = LD^2 x 2^((SPL-SS)/3), where LD is listening distance in m, SPL is peak dB per channel (105 in cinemas), SS is each speaker's sensitivity spec in SPL/1w/1m. Then double it for a 3dB margin of safety.

[59] An example of disinformation is the subject of subwoofers that divides audiophiles into two camps: *for* & *agin'*.

Gallery – "A picture is worth a thousand words."

"The two things that really drew me to vinyl were the expense and the inconvenience."

The New Yorker

Courtesy AudioScienceReview.com, among preamps ~$200 with good SINAD (signal to noise and distortion), an ART DJ-Pre-II has two C-load choices but no balance, and low input overload. Emotive XPS-1 has no C-load or balance. Cambridge Audio Solo has balance, but no C-load, and RIAA ±0.65dB. Schitt Mani2 has two C-load, two R-load choices, but no balance. All these essential controls are provided in the maker preamp project **p106**.

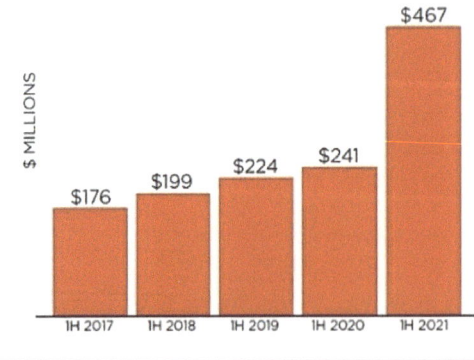

2021 record revenues (not incl. private sales).

Above: South Rampart St Ramblers record the first publicly issued stereo album on a Cook twin mono track LP.
Below L: Emory Cook released 75 binaural albums 1952~58. ***R:*** The author's way around Cook's twin-tonearm.

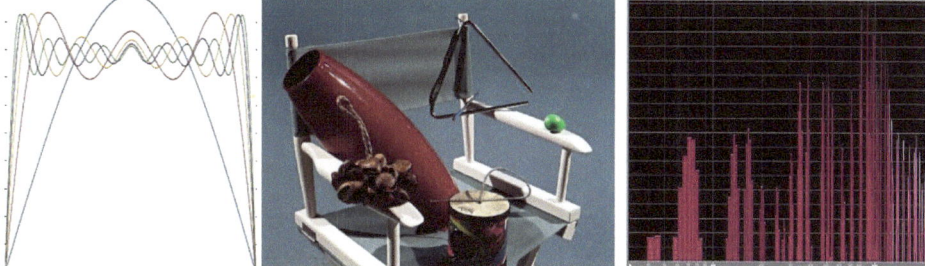

L: Digital clipping (squaring) produces >100 odd-harmonics, here adding 3x, 5x, 7x, 9x to the original frequency [Wilbraham, Gibbs]. All sounds are an instantaneous mix of sinusoidal waves, usually odd only or even & odd multiples. ***Middle:*** "Non-sentimental" sounds avoid the bias listeners have for music genre, used for subjective blind evaluation of loudspeakers by comparing them live to high-quality recordings using the microphone **p94**.
R: Rising sharply in frequency, a spectrograph of the infamous orchestral triangle (in middle image), the ultimate test of any audio device. Few pass without a distinct "clunk" of *transient intermodulation distortion* (TIM).

Gallery – "A picture is worth a thousand words."

Frequency response in grooves mastered by a WE *rubber line* lathe in countless 78s. Quite flat 50~6kHz, an octave beyond the telephone and just better than AM radio of its day. The 33⅓ LP added the top octave+.

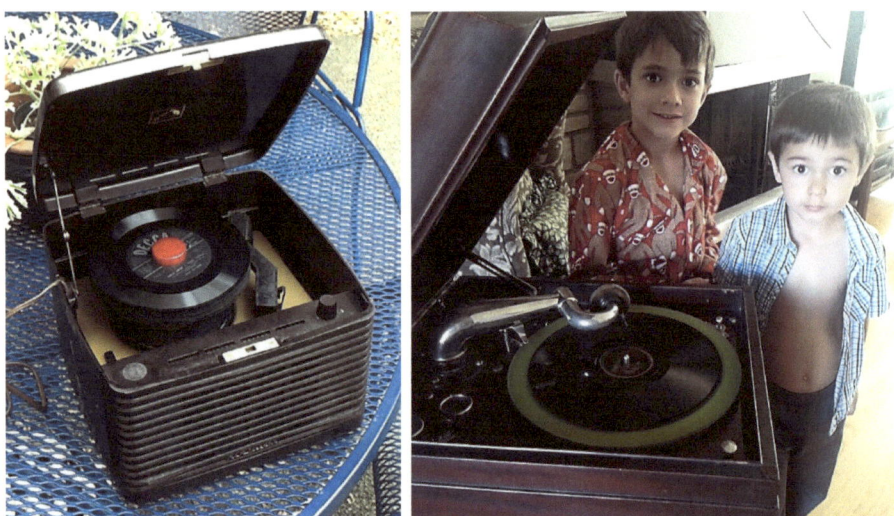

L: RCA's wildly popular 45rpm portable record changer with an awful sounding crystal pickup. *R:* Author's grandsons mesmerized by a 100+ year-old wind-up Victrola IX playing an acoustically recorded 78 shellac.

L: The author's rig for stylus examination and microphotography up to 800x for this book. *R:* In Flaherty's film "Nanook of the North" 1922, the Inuit Canadian is wowed by even the primitive reproduction of his own voice.

A maker phono preamplifier with essential controls

Here is the "inside story" of a "phono stage" (phonograph preamplifier; semi-technical). Modifications at a beginner level transform an inexpensive pre-assembled & tested printed circuit board (PCB) for ~$19 including shipping, shown **below** as received before modding. Work requires only basic hand tools [60] and parts that follow. User provides 12~30vdc wall Power Supply (PS) using a *RailSplitter,* or bipolar ±6~15vdc, and optional box. (Off-shelf preamps for $2~300 are in the Gallery section, but none has all the controls in this project.)

The modified preamp [61] [based on Lipschitz & Jung] is for stereo moving magnet\iron (MM\MI) pickup cartridge to *optimize reproduction timbre (tone color) for both monaural and stereo* 33⅓ or 45rpm disks, or 78rpm releases using "re-EQ" to convert from RIAA. [62]

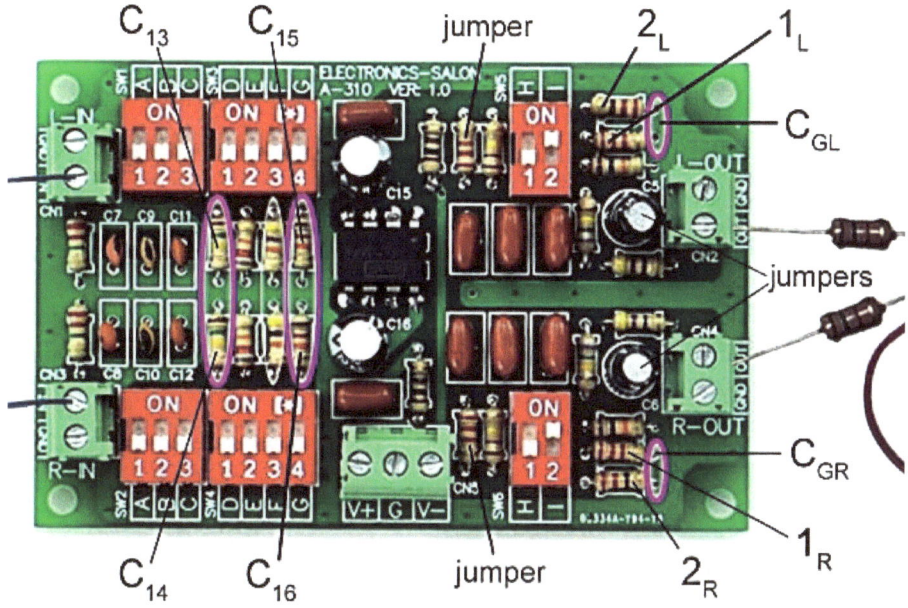

Acquire the following components...

> PCB #A-310 from http://stores.ebay.com/Electronics-Salon or
> https://www.aliexpress.com/i/2251832427881883.html?gatewayAdapt=4itemAdapt
>
> *Resistors (6):* 1% film Mouser#271-**732**-RC, 271-**825**-RC (2), 271-**910**-RC, 271-**1.65K**-RC (2).
>
> *Capacitors (6):* 647-UES1E**220**MEM (2); 505-MKP20.**22**/100/5 (2); 594-K**221**J15C0GH5TH5 (2).
>
> *Switch:* SPST audio Mouser#506-MTA106D. *Inductors:* Fe beads Mouser#623-2643021801 (2).
>
> *Hardware:* phono jacks (4); ~#28 solid hook-up wire; ~#22 stranded in red & black. Metal box?
>
> *Recommended active part:* "RailSplitter" Mouser#595-TLE2426CLPR. [in UK http://uk.rs-online.com/web/p/voltage-references/0284220/ New Zealand http://nz.rs-online.com/web/

---

[60] *Tools:* pliers, soldering iron, solder-wick, pin vice w/#70 bit. Mono disk for balancing channels. Optional ACVM to set levels (workaround below); capacitance meter to measure C_{cable} and to match capacitors.

[61] IP FOR PERSONAL USE ONLY. USE FOR SALE OR PROFIT ONLY BY WRITTEN AGREEMENT WITH FILMAKER TECHNOLOGY.

[62] Product of https://www.esotericsound.com/. Beyond the scope of this paper are the laudable merits of other preamp topologies [Holman, etc.], or beyond basic library preservation\restoration techniques [Copeland, etc.].

The purpose of modifications below are for: *a)* precise balance for cartridge sensitivity errors of up to ±2½dB to eliminate vertical distortion playing mono disks, and to improve soundstage for stereo; *b)* VLF stability, reduce noise & distortion; *c)* add selections for capacitance-cartridge "tuning;" *d)* & *e)* improve frequency response to 30~20kHz ±¼dB (typical) for accurate timbre (tone color) and phase response for unsmeared transients. Modification time is 1.5~3 hours for the 12 part substitutions and eliminating four others.

Modify an A-310 PCB in five (5) check-by-steps

a) ☐ In image **above**, remove one at a time four resistors 1_L, 2_L, 2_L, & 2_R by heating each right pad, prying out its lead, then its left lead. Still heating the *left* pad, insert & solder only the *left* lead of new resistors $1_L=825\Omega$, $2_L=910\Omega$, $1_R=825\Omega$, & $2_R=732\Omega$.

b) ☐ In a)'s freed holes within each **violet oval** at "C_{GL}" & "C_{GR}," heat each pad to insert and solder one lead of each **22μF** non-polarized, low ESR electrolytic, plus one lead of each **220nF** poly capacitors as **below** and **overleaf**. Connect the two free leads plus two free right leads of a)'s resistor pairs. Physically arrange and stabilize each set of four components, then solder the "flying junction" of four (4) leads each.

c) ☐ Directly off the IC's left (ignoring PCB labels), remove the nearest two resistors and substitute re the 1st illustration C15 & C16, **220pF** ceramic discs (NPO or COG type, 5 or 10% tolerance). [For C-loads 470~859pF, substitute **470pF** at C13 & C14.]

Reworking at a PCB solder pad may require heating it through solder braid (or a wig of fine stranded wires) to sop up excess solder. If pad hole is not opened, drill it out using the pin vice w/#70 bit. Take care not to overheat, or cause pad to disconnect from its PCB traces.

A310 mod1ii PCB top   R Miller FilmakerTechnology

For the 12 component substitutions, eliminating four others 1.5~3 hours. See closeups **opposite**.

L: "flying junction" with new electro-lytic laid left, polypropylene right. *R:* "RailSplitter" powering.

d) ☐ Referring to the illustration **below** of the PCB's underside, in **red** ovals nearer the middle, ***tack-solder*** jumper wires to short top-labeled resistors R22 & R24.

e) ☐ In **red** ovals at left, solder a ***bridge*** to short capacitors C5 & C6. *You're done!*

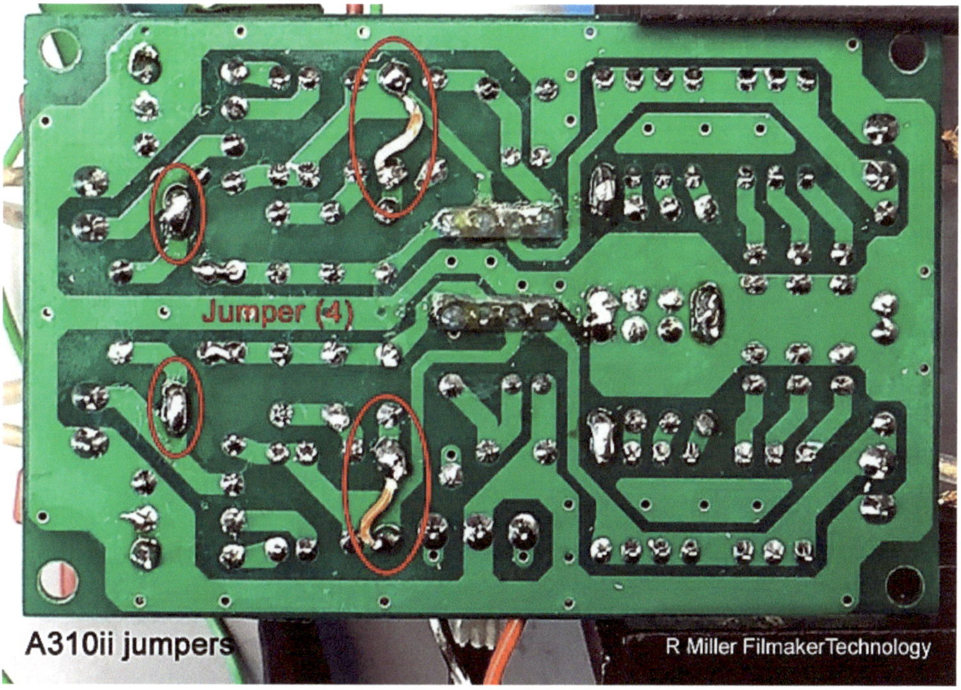

PCB underside showing four jumpers, two by solder bridges (blobs) and two by wires.

A maker phono preamplifier with essential controls

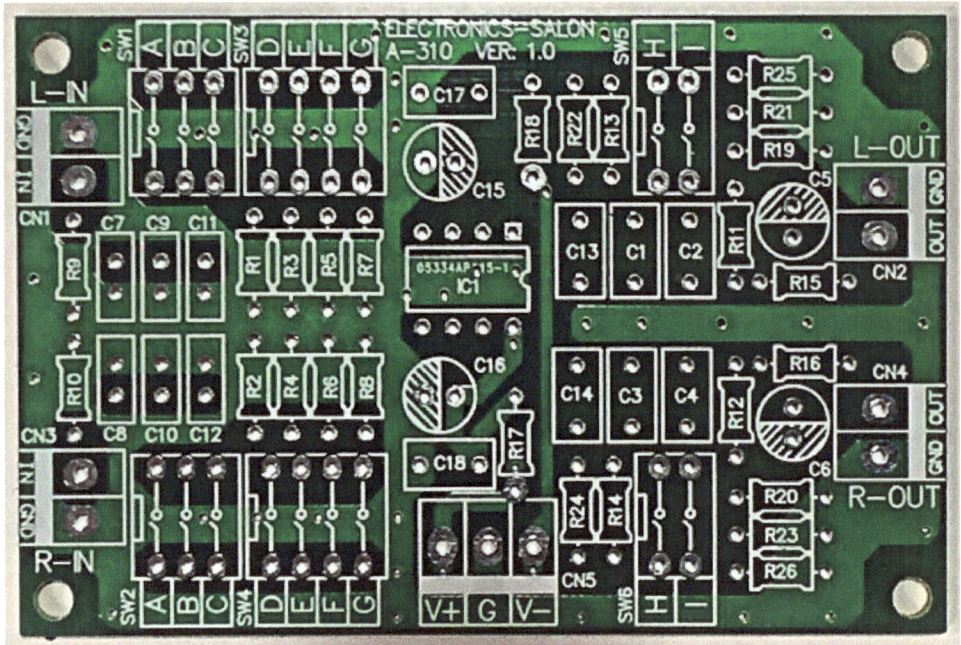

An unpopulated PCB shows component labels. R21, 25, 23, & 26 are substituted in step a); R7 & 8 in step c).

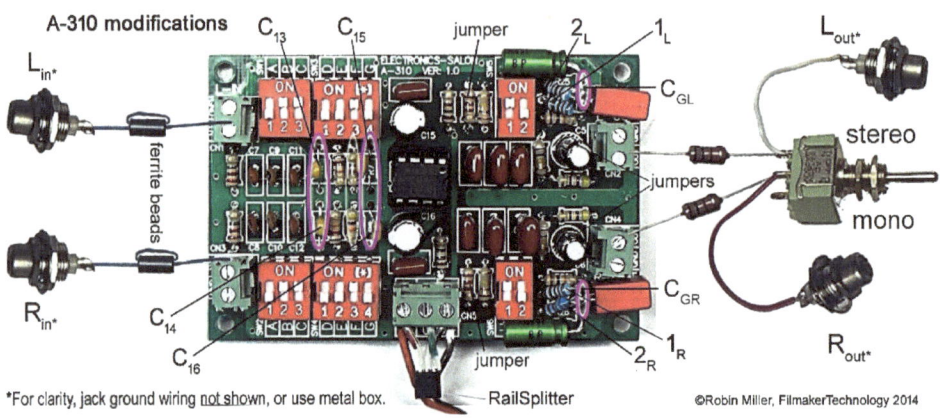

*For clarity, jack ground wiring not shown, or use metal box. —RailSplitter   ©Robin Miller, FilmakerTechnology 2014

The PCB's original documentation no longer pertains - refer only to this chapter.

Wiring, setting up, and using the modified preamp

Once PCB modifications are done, time to wire to\from the screw terminals by referring to the illustration **above**. *If not fitted in a metal box, add ground wires between the outer shells of each of four (4) jacks and the PCB "G" terminal. Ground wires are not needed if jacks are mounted in a non-insulating metal box, which is then connected with one wire to PCB "G." For the later procedure to compensate cartridge imbalance, make input\output wires as short as possible, but long enough to reach the other channel's terminals.*

Wire the PCB "IN+" terminals through a ferrite bead, loop around, through again, then to the center pin lug of the RCA jacks. Four (4) signal conductors from the cartridge outputs (natively balanced) connect via 5ft foot (1.5m) maximum of low capacitance 25pF/ft coax. Better is using shielded\unshielded twisted pairs (STP\UTP, even CATx LAN cable).

Wire the tonearm ground conductor to the PCB power center terminal. If the cartridge has one – pin strapped to its case, remove if the metal body is not insulated from the arm.

Each L & R "OUT" is wired through a 1,650Ω resistor to its output jack's center pin, and to a single pole switch between them to "mono" a lateral groove. *For vertical hill-and-dale recordings, temporarily reverse R+ and R– clips at the cartridge (gain settings remain).*

Select cartridge loading using the cart loading table below for switch combinations A~G. C- and R-load switches A~G may be engaged in any combination *per your cartridge's specifications (same setting for L & R), including the tone-arm-to-preamp wiring capacitance as discussed.* [63] **Later will be balancing the cartridge's sensitivity.**

A-310 R-load modif			A-310 C-load modif		
switches ON	R (Ω)	D* (150k)	switches ON	C (pF)	D* (470pF)
none	140k*		none	0	
A	82k	E	A	22	
B	52k	DE	B	47	BD
AB	46k	F	AB	69	CD
C	29k	EF	C	100	ABCD
AC	25k	DEF	AC	122	DG
			BC	147	
opt'l C-load modif			ABC	169	
C (pF)			G	220	
470*			AG	242	
517			BG	267	ABDG
570			ABG	289	
639			CG	320	
690			ACG	342	ABCDG
759			ABCG	389	
859					

*D choice

©Robin Miller, Filmaker Technology

R & C cart loading table for setting switch combinations A~G on the PCB.

Connect the PCB to external equipment

Check off the following steps as you perform them in order:

☐ Connect turntable signal cables to the left L and right R inputs of the PCB. Connect tonearm ground lead to ground screw. (Make sure any cartridge ground strap is removed.)

☐ Connect the outputs following the "mono" switch to the stereo receiver/amplifier. If not mounted in a metal enclosure, take care not to stress the switch's two flying resistors.

☐ Connect a bi-polar power up to +\–15vdc. Or for no appreciable DC at the outputs, use a 12~30vdc power adapter and the RailSplitter shown to create virtual bipolar power.

☐ Set for both channels the cart loading R & C switches per manufacturer's spec minus the C of tonearm interconnects (measured or estimated). See loading table above. The conventional R-load of 47kΩ requires only switch "F" on ("D" & "E" are off).

---

[63] If no C-meter is available, approximate pF at 25 times its length in feet, including the distance from cart through the tonearm. E.g. if 5ft long and the cartridge requires 275pF, set ON switches B & C.

Wiring, setting up, and using the modified preamp

☐ Apply power to the PCB preamplifier, set the tonearm on a stereo disk, and listen for *spatial* sound from both channels.[64] (If hum is heard, see "Hum…" in paragraphs below.)

☐ After setting the gain balancing switches per the "Tuning..." section, enjoy the music, stereo or mono. Keeping the bare PCB free of dust accumulation is the only maintenance.

Pickup signals are at a low millivolt level, and at medium-high impedance, so input wiring should be kept short, using either twisted wire pairs or low-capacitance coaxial cable (~25pF/ft). Raised in level up to 60dB, the preamp's output is then "IHF line level" of ~300mV, and low impedance, so wiring may be 10+ft (3+m) of coax, longer if shielded 2-conductor twisted pair cable is used, the shield connected at only one end. (In high a RF environment, such as near a radio transmitter, connect the other end of the shield to ground through a 0.02µF (20nF) capacitor.) The destination input impedance should be ≥20kΩ.

Three ways to power the preamp

Power the preamp 1) with a regulated *bipolar supply* up to the OPA2134's absolute maximum ±18Vdc to provide additional headroom, eliminating an over-modulated 45's, or pops from dust latching up the IC that requires a power cycle. Or 2) use two to four 9V batteries, wired through a 2-pole switch, giving about 100hr of use per set of fresh batteries.

Perhaps best, as shown on **p111,** is to 3) use Analog Devices' TLE2426 "RailSplitter" and almost any 12~30volt *DC* "wall wart" in your spares box, even if unregulated. The PCB has power filtering including RF, so the power supply may be up to several feet away.

USING A RAILSPLITTER, DO **NOT** POWER ANY OTHER COMPONENT IN THE AUDIO CHAIN WITH THE SAME POWER ADAPTER, AS SHORTING OF THE **PCB** CAN RESULT.

After powering to avoid hum, the final step is *balancing* cartridge channel differences.

Avoiding hum by design

Unless the preamp is mounted in error too close (less than 1ft) from a turntable's AC motor, amplifier power transformer, or any AC main carrying high current, even the bare PCB is very likely NOT the source of hum. The fault lies in how it is connected to the outside world, inadvertently creating a "ground-loop." Eliminating hum is therefore the user's responsibility to implement hum-free wiring. It is as simple as breaking the *loop*.

When *one ground only* is connected, a *ground loop* cannot complete the circuit that introduces hum. One cause of hum is employing BOTH the cart's strap from its case to a minus output pin, AND the ground lead from the tonearm & headshell to the PCB. Another cause is grounding both ends of any shield, including allowing any bare shield to come into contact with the metal tonearm – the shield should be grounded only at one point. Best is to use only the tonearm ground lead, and remove any signal pin ground strap. The cartridge body is grounded only by being bonded (electrically connected) to the headshell\tonearm.[65]

A common cause of hum is an external component, (a receiver/amplifier or the PCB power supply) faulty in design from a "pin 1 problem," that permits a loop through the AC mains ground system. This may affect other components of the audio system as well. Lift a mains ground pin to isolate the problem unit, but this is an unsafe beyond a brief test.

---

[64] If any channel produces no sound, power OFF, recheck all connections, and check receiver/amplifier settings.

[65] Because phonograph cartridge coils are inherently floating sources, a preamp with balanced inputs cancels induced hum. However most preamps are unbalanced: dress <6ft (<2m) interconnects away from hum sources.

How to balance a cartridge's channels with a modified A-310 phono stage

Either L or R gain switches #1 and #2 are never both ON together. To balance your stereo cartridge's channel sensitivity difference (if not more than a defective 2½dB):

1) Begin by setting both channels' gain switches to #1 ON, #2 OFF. *While playing a mono record with the mono mode switch in "stereo,"* observe the output levels and determine the difference. If the channels differ by less than ½dB, skip to step 5).

   (Tip using no meter: reverse the polarity of the wires of one cartridge channel, place the preamp switch in mono, and listen for the lowest level as signals cancel.)

2) If the system's L-channel is <u>higher</u> than R by more than ½dB, skip to step 3). But if <u>lower</u>, reverse each pair of inputs & outputs at the PCB so its labeled "L" side is in the system's R-channel, and the board's "R" side is in system L. Continue steps 3) & 4), but *address the #1 & #2 switches and its channel written {L} on the opposite side of the PCB labeled "R," and those written {R} on PCB side "L."*

3) Now for the system's {L}-channel level being <u>higher</u> than the {R}-channel by more than ½dB, flip the PCB's {L}-channel #1 switch OFF, #2 switch ON, and again observe the level difference. If less than ½dB, skip to step 5).

4) If system's {L}-channel is <u>still higher</u> than {R} by more than ½dB, switch PCB {R}-channel #1 OFF, #2 ON, and again observe the level difference. It should now be less than ½dB – go to step 5).

   If not ≤½dB, then the cartridge (or level metering) is defective, and you cannot proceed until one or more are corrected. (Evaluate metering by swapping channels and note any disparity in readings, or measure using only one meter.)

5) Your system is balanced within less than ½dB, which provides at least 24dB (>94%) reduction of vertical artifacts. Switch to "mono" for cleaner sound playing lateral monaural recordings. And enjoy a finely balanced soundstage for stereo recordings.

A UK reader's modified A310 nicely mounted in a metal enclosure, working great after C-load selection and gain balancing. (Or prewired & grommeted for an edge-slotted metal enclosure without unwiring.)

Wiring, setting up, and using the modified preamp

Measured performance of the modified preamplifier

The modified single-stage RIAA preamplifier, offering simplicity and low cost, is by Lipshitz [1979]. Unmodified, undesirable VHF boost occurs by the op-amps' unity gain; distortion by audio-carrying capacitors. How the modified preamp's frequency response with design simulations is **overleaf**. You can pay more than $35 plus power supply & case, if any. If not pretty (or expensive) enough to call attention to, hide the bare PCB under the turntable near the arm post, the arm wires secured directly to the board's "IN" terminals. (Compensate for the shorter cables, now reduced ~100pF, with a higher C-load selection.)

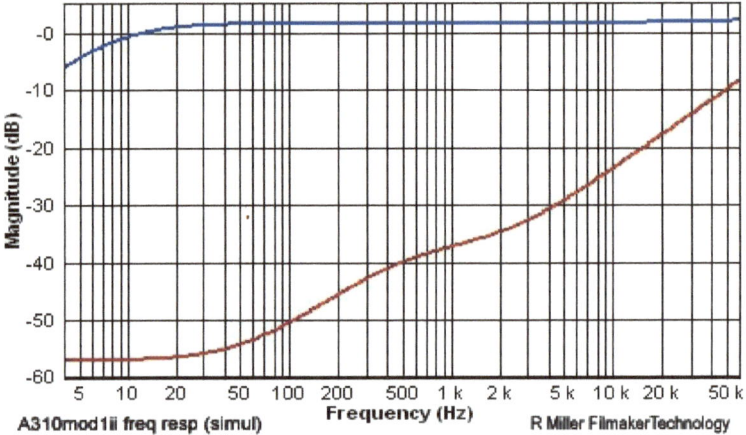

The author's design simulations show a near ideal response (blue) for accurate timbre. As input to the preamp (red) is a simulated ideal RIAA groove as mastered & cartridge.

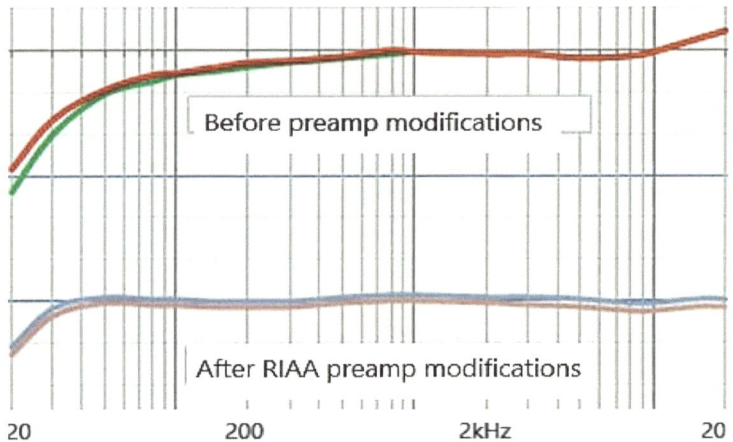

Frequency response at 1dB per vertical division of the original A-310 (red & green) and with too low gain of 30dB v. the modified preamp (pink & blue separated 1/10dB for clarity), ±<¼dB 40~20kHz, -1dB at 20Hz, and with useful gain of 39dB that agrees well with its simulation, above.

In the chart on the previous page, the original frequency response (with channels graphed separating 1/10dB for clarity) reveals it will have thin bass and emphasized clicks & pops. After modification, a measured ±<¼dB 40~20kHz (–1dB at 20Hz) shows the frequency and phase responses, desirable for precisely decoding the modulations mastered in the groove.

Easy component changes make for greatly improved audio performance, and with 9dB more useful gain for MI\MM cartridges of 38, 39, or 40dB. The modifications offer a wide selection of cartridge capacitive loads that can realize flattest response for any cartridge.

Along with accurate (timbre) tone color, the holy grail of high fidelity, transients are restored and audibly improved. Adjustable balance optimizes mono quality by nulling distortion, and optimizing the stereo soundstage. It all means more music enjoyment.

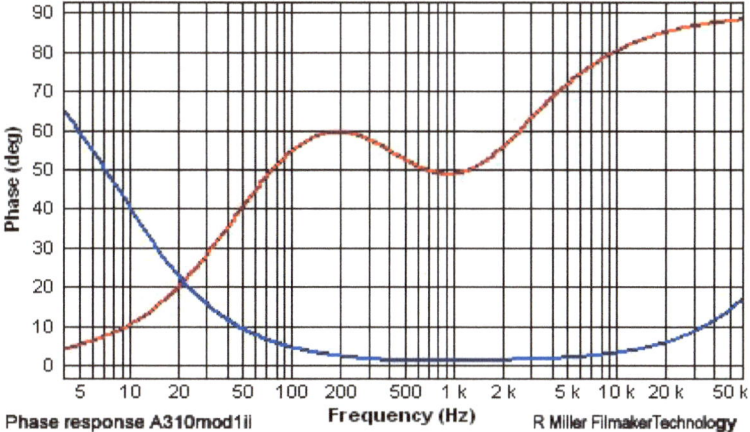

Phase response of the inverse RIAA filters used in disk mastering is baked in the LP groove (red). The modified preamp design reconstructs transients for a very good round trip phase error <5° 100~15kHz (blue).

Installing the PCB, connectors, and mono switch in an enclosure...

Clicks and pops even when not playing a record are a symptom of radio frequency interference (RFI), usually resolved by a metal enclosure (not plastic). An aluminum or steel box comfortably larger than the PCB is drilled for enclosing the preamp and jacks. [66] Firstly, consider where the box will be located: sitting on a shelf near the tonearm corner of the turntable, or attached under the turntable base (plinth). Before marking the box for drilling seven holes\slots, sketch your plan with dimensions. Envision which surface of the box will be the mounting side, and whether holes will be needed for mounting screws. Envision the PCB inside, and how it can be secure, maybe only by non-conductive putty.

Envision three (3) holes for the input jacks and ground screw, inserted from the inside. Envision room for wires behind these. Envision three (3) holes for the output jacks and the stereo-mono switch. Mark the drilling locations, with the diameter drill for each. (Sharpie marks can be scrubbed using cotton swabs and alcohol.) Envision a hole for a connector or grommet (1) for the power cable (and audio cables not using jacks). If not disassembling a prewired PCB, locate those holes near an edge for slots cut through with a metal shears, saw, or nibbling tool to slide in grommeted externals. Measure twice, drill once!

a) Input jacks may be either insulated, with two wires from each jack to its side of the PCB), or, using smaller holes without jack insulators, electrically bonded to the metal box with only the center wire connected to "L/R" of the PCB input screw terminals.

b) Tightly "bond" the ground screw electrically to the box between lock washers (for the tonearm's ground lead/spade lug, secured under a wingnut outside). Inside connect to the "ground-zero" center position (common) of the PCB's power screw terminal.

c) If using nibbled slots, slide into position the jacks, switch, or any grommet(s) without the need to disassemble a pre-wired PCB.

---

[66] A Hammond extruded box (Mouser #546-1455J1201) is one example. Drill a total of seven (7) holes\slots.

d) Secure the mono switch. (Cancels vertical artifacts for lowest distortion mono.)

e) Attach the box's cover or split sections, careful not to pinch any wires. (Normally no cooling vent holes will be needed.) Apply adhesive rubber feet if desired

f) In an aesthetically acceptable manner using printed labels, transfers, or a simple Sharpie, label L & R inputs, ground lug, L & R outputs, stereo-mono switch, and power connections with maximum voltage(s). Also label that user-selectable R/C-load and channel balancing switches are inside for any change of cartridge or tonearm wiring. The Load & Gain switch table below can be handier if pasted inside the box.

Conclusion installing the modified RIAA phono stage

Despite its simplicity and low cost, the modified A310 printed circuit board (PCB) is a remarkably high-performing preamplifier for moving magnet (MM) \ moving iron (MI) pickups (cartridges) having a nominal output of 3.5mV/cm stereo (5mV/cm mono).

Critical are the capacitive and resistive loads, switch-selectable on board to match any pickup's specifications in order to realize the manufacturer's claimed frequency responses playing RIAA standard records. That quality will then be passed faithfully along the audio chain by the modified preamp's accurate frequency & phase responses and low distortion. Delivering lifelike percussive transients, guitar pluck, and woodwind *chiff*. The true timbre (tone color) that is the *sonic character* of each instrument or voice.

Timbre is as recognizable as a familiar face, and to ears that with experience listening become discerning, even individual instruments' sounds can be as unique as fingerprints. Distortion is like applying face makeup that ranges from too much to plastic surgery! Attention-getting, but phony. Unlike preserving the intentions of musicians and producers, purely subjective listener "preferences" (opinions) are fleeting by trial & error, often are false tone color, and subject to change, before or after buying a new part. Reading online audio forums or FaceBook groups, one observes that audiophile opinions often disagree.

The prior "**Tuning…**" chapter reveals the extent to which, for decades, ignorance regarding cartridge loading has caused bogus opinions by trade magazine reviewers and audiophiles, such as proclaiming that a pickup sounds "dull" or "brittle;" bass relatively "thin" or "thick." C-load is needed to *complete* the cartridge's circuitry and performance, but are beyond the control of the cartridge manufacturer, thus are user-implemented. Often proper loading is not changed as reviewers and audiophiles "roll" (interchange) cartridges. Such oversights lead to invalid judgments, because uncontrolled variables are still in play.

Using a fast, high impedance JFET opamp, the modified preamp effects precise RIAA equalization with a flattened magnitude-frequency response – see the chart on **p116**. The resulting phase response on **p117** completes precisely compensating the inverse RIAA-made master recording – after proper R- & C-loading of the replay phono cartridge.

For best stereo "soundstage" and cleanest monophonic reproduction, gain adjustments balance the cart's outputs within less than ½dB (typically ±¼dB), selected using switches for each channel on the circuit board. Power required is bipolar ±6 to ±15vdc (by a "wall-wart" or batteries), but a single 12~30vdc supply and a "rail splitter" is the best method.

The preamp may be customized to suit the user, especially whether the user will provide an enclosure, and where the preamp will reside. One author-prepared assembly is on **p115**, with all interconnects for using the PCB bare or boxed. It anticipates a user upgrading anytime to an enclosure with slots from edges for sliding in connectors, grommeted wires, and stereo-mono switch without disassembling the wired & tested configuration shown.

A transcription tonearm – how come; how-to

We have seen the steampunk tonearm earlier, on **p83~85**." I've made several; you might want to *be a maker* of one too. But *how-come?*

It began a long time ago in a place far away: "Electrical Transcriptions" (ETs) were large 16in records, not for sale to the public, used by syndicators to distribute radio programs to stations. In the day when home disk recordings were limited to 3~4½ minutes, entire 15, 30, & 60minute shows were aired from 1 to 4 disks. And filled most local stations' airtime. Advertiser sponsored, they featured top-quality, live-sounding production with a host\announcer, celebrity performer(s), orchestra, and the approving sounds of a "live" (on disk) audience. To prevent listeners storming their local station, thinking the celebrities were right across town, the FCC mandated the station announce: "The following program is *transcribed.*" Thousands of these large ETs were mailed every week for the purpose, not surprisingly, of hawking their sponsors' messages. [More radio history in the author's book "American Radio Then & Now: stories of Local Radio from The Golden Age," available at Amazon\Kindle.] 16in transcription disks required large turntables 16in in diameter – and tonearms that could reach across them. But today these "transcription tonearms" have another quality: they produce the lowest *tracking distortion* for any size disk.

A typical "8~9 inch" (200~230mm) pivoted tonearm adds up to 1% distortion solely due to its length. The "12 inch" (~305mm) arm described in this chapter produces at most half that distortion. For this reason, well-heeled audiophiles pay $600~$6,000 and more for a 12incher (Americans call them "16in" for their disk size limit, if not their overall length). They might add a 2nd, 3rd (or 4th?) arm to handle more conveniently a mix of formats! For the project below, the parts are available at any local hardware store for about $35 – same almost trivial cost as the high-performing preamp in previous chapters. Maybe free in your junk drawer?

So for ~$300 total, a newbie to vinyl can find for around $150 if not $10 at a yard sale a used turntable from eBay or Good Will Industries, add the DIY $35 preamp, the DIY $35 tonearm, a good cartridge & stylus, and have enough left for a trip to the new\used\online record store for a few dozen disks starting collection. *$300 and aligning to get better sound from vinyl!*

This next chapter gives the instructions for making yourself, as well as properly aligning, a high-performing transcription tonearm. What it lacks in universally acclaimed beauty it makes up for in a modest ingenuity – the author's *and yours.* It requires moderate skills with a drill press, a tap & die set, and common hand tools. Unhurried, it takes about a day to machine and assemble the arm once materials are at hand. There might be blood, so have a first-aid kit handy. This pain and effort will prove a small price to pay for many years of enjoyment playing vinyl with a nicely handling arm.

The alignment is universal for any size disk, optimized for 0.47% maximum distortion tracking 12in LPs. Add it as a second (third?) arm to an existing setup. And align them optimally for specific size disks, with styli for different groove widths. (The image below shows a 1950s broadcast turntable with the requisite two 12in arms and four disk sizes piled on.)

An aside: Now at ¾ through this book, you must have wondered about the OCD of the author. [I prefer CDO, the initials in alphabetical order.] And whether, with your own hopefully milder case, you would rather just spend $300 on a brand-new turntable with an integrated tonearm and serviceable cartridge and be done with it. I understand. For context, do you know any model railroaders? [I am speaking from experience; my father *was* Miller Backyard Railroads.] The present hobby is as rewarding, is less expensive, and doesn't require help to lift anything. As with the DIY preamp, the results will return your investment in effort for years. And both projects would be hard to equal at any price. *So hang in!*

Popular with broadcasters were two or more of this restored RCA model BQ-2B 16-inch turntable. The 12in arm at the right is aligned for 7, 10 & 12in disks; the 12in arm at back for 16in ETs; all four sizes shown sharing the spindle. The 16in format enabled 15min *transcribed* programs, or 30min and hour-long shows by *segueing* between two turntables.

"Steampunk" 12in transcription tonearm

Abstract: For a quick solution only to evaluate a rehabilitated 1940s 16in turntable, I made this arm with ordinary hardware. The resulting low mass 12in (305mm) tonearm worked and sounded far better than expected, tracking as low as 1¼g. It led the author to greater appreciation of the evolution over more than a century and a half of the analog phonograph, which is enjoying a comeback. For hobbyists, here follows maker instructions for a rewarding project for near $0.

1. Background & purpose - a high-performing 12in (305mm) phonograph arm

Even if you don't have, or plan to acquire, any 16in diameter "electrical transcription" (ET) disks, and 16in broadcast turntable to play them, your vinyl/shellac record collection can benefit from a long tonearm. In the US sometimes termed "16in" (for their overall length), they typically are ~12in (305mm) in effective length, tonearm pivot to stylus tip. More common are shorter ~9in (225mm) arms for LPs. Longer 12in "transcription-length" tonearms imply lower distortion, lower tip & disk wear, and just *feel professional.*

With a smaller offset angle, longer arms do not suffer as much "skating" (inward force due to friction) or its variability disk to disk. With their shorter overhang, they have lower distortion due to tracking error, where the stylus cantilever is not tangent to groove. 12+in tonearms were/are accessories for disk mastering, broadcasting, & archiving, implying high reproduction quality. Many professional and a few hobbyist turntables have two or more tonearms to accommodate different disk formats and sizes. It may require mounting to an outboard *arm-board* for an existing turntable, a preamp input switch, or a second preamp.

The Steampunk 12in tonearm alongside a 8½in arm. Although intended for 16in disks on 16in turntables, they have advantages for 12in records on 12in turntables. *Never mind it's not pretty.*

Underlying this project is the history and realities of acoustical-electro-mechanical phonograph reproduction: We learn to appreciate the tremendous development over nearly a century and a half, at first as a dictation device, and its science in service of spoken and musical art-forms. Second, we must acknowledge that technically superior methods have succeeded it: convenient and editable magnetic tape and superior digital sampling. Yet the phonograph evolved to offer competitive audio quality in mass distribution. Third, notwithstanding great strides, no audio reproduction equals real hearing, verisimilitude, that requires full-sphere 3D recording & reproduction – for papers about *Ambisonics* [Gerzon] and "High Sonic Definition HSD-3D" see www.filmaker.com/papers.htm. Yet well-made phonograph records from the *hi-fi* era on can and often do sound better than one expects. Fourth, the century and a half of recorded history is preserved mostly on analog records!

My first job, not working for my dad, was at a 250-watt #70 market radio station, its 16in turntables having 12in tonearms. [67] They were reliable (never skipped) and sounded great – at the time. Along with their industrial Art Deco aesthetic, handling them felt good; back-cueing and slip-cueing, they just seemed 'right.' In those days on Radio, they gave a 17yr old a motivating sense of professionalism. The peer reviewers of this book agree – most also started in Radio. Today, transcription tonearms can add not only to higher audio quality, but to the *experience* of playing vinyl. And the icing on that cake is making one.

In the Golden Age of Radio from the late 1920s through the 60s, program syndicators mailed to local stations and the Armed Forces Radio Network sponsored programs by the thousands recorded on 16-inch ETs (LPs are 12in, some 10in). Although at 33⅓rpm, ETs have 2mil (50μm) grooves, a compromise 2.5~2.7mil (63~69μm) stylus is used to play ETs and the 3mil (75μm) coarse grooves of the Standard Play (SP) 78rpm disks. ETs were supposed destroyed after two airings, but many survive today, bearing drama and musical variety, with stars such as Sinatra & Fitzgerald, and Basie & Ellington big bands of the era.

Even after introduction in 1948 of the 12in microgroove LP that shrunk the 78rpm's 3mil stylus to 1mil (25μm) across, broadcasters continued to use their indestructible 16in turntables with later, lighter tracking, and even better handling and audio quality 12in arms. Today as an audio engineer and sound conservator, I've restored and use several turntables with five (5) 12+in tonearms playing 7~16in disks. These transcription arms still feel *right*.

Compared to light 8~9in arms, 12in arms have higher "mass" (moment of inertia), matched by middle compliance (springiness) cantilevers. Disk mastering engineers used 12in (305mm) arms to span the 14in lacquer blanks they used to master 12in LPs. Pickup cartridges far outlast styli that eventually wear out, so cartridges accepting interchangeable styli make the most sense to these professionals. Although "new" styli are often inferior knockoffs, and OEM new-old-stock (NOS) are disappearing. With a still useable cantilever & elastomer, Expert Stylus UK and others offer the service to *re-tip* a stylus' cantilever.

Expert Stylus says the wear is "safe" up to 3 grams vertical tracking force (VTF). At this pressure, the DIY arm described here can mate with medium compliance styli, 10~20CU. [68] Most brands have comparable selections, too many to list here, but findable online, MC included. With longer arms, skating is less, and more immune to differing vinyl friction.

With its adjustable mass, the Steampunk arm can also accommodate higher compliance 18~25CU styli, along with Ortofon, Shure, and Audio Technica, etc.) tracking 1¼~3g. True, 8~9in (200~230mm) arms take less room, but they require 40~60% more anti-skating compensation, are more prone to hop grooves, and inherently generate 50% more tracking

---

[67] Also used as turntables for Vitaphone motion picture talkies with soundtracks on 16in disks.

[68] E.g. a Pickering V15 or Stanton 500 with 10CU D5100A(L) spherical stylus, and 12CU D5100E(L) 0.4mil or EE 0.3mil ellipticals; a Pickering XV15 or Stanton 680\681 with D6800EL elliptical or Stereohedron D6800SL...

distortion than 12in arms. A 12in transcription tonearm is a wise choice for its universal adaptability switching between record formats, as well as for better sound.

The steps for making an experimental 12in transcription tonearm follow.[69] Its Spartan "industrial design" & construction are simplified as much as possible. You need a modest shop's typical hand tools (pliers, screwdrivers, hacksaw), a vise, a drill press (or portable drill press adapter), file, emery cloth, and a 30W soldering pencil. You also need a tap & die set with a ⅜x16 thread tap. Cost: a day or so of your time, but ~$35 US in parts – less if you have them already, as I did for the prototype, costing $0! The project is engineered so the precision of parts and machining is forgiving. Modest craftsmanship is acceptable, except items marked * that require care. The *help-line* are FAQs refreshed as needed at www.filmaker.com/papers/UPDATES_RMiller-Better Sound of the Phonograph.pdf . But I cannot warrant your results or be responsible for errors, omissions, or collateral damage.

The contents of this chapter's maker instructions are organized in nine (9) sections:

1. *Introduction & purpose;*
2. *Bill of materials and initial machining;*
3. *Fabricate wand sub-assembly;*
4. *Wiring (up to preamp connectors, left to user's choice);*
5. *Fabricate finger lift, armrest & clamp sub-assembly;*
6. *Tonearm installation on base, alignment, & testing;*
7. *Fine tuning – adjust arm resonance, preamp interfacing;*
8. *Conclusion – further considerations, and what to expect?*
9. *Addenda – pictorial tips, reports in use, 2nd & 3rd prototypes.*

Sections §2~5 cover construction of the arm; §6 installing it on a turntable. §7 & 8 address performance of the arm + $35 DIY preamp of the previous chapter that, with a low rumble turntable, complete a "high-end" DIY turntable. §9 adds pictorial tips, including a DIY VTF gauge. *For the benefit of familiarity, scan the instructions to p115 before beginning.*

2. Bill of materials and initial machining *(check off boxes as you go)*

- ☐ *½x½ x 1/16in-gauge aluminum angle, cut precisely 16in long. **[1in = 25.4mm]**
- ☐ ⅜-16 zinc-plated steel all-thread 2¾in long. (Or any metal substitute.)
- ☐ ⅜-16 "coupler" ⅝x1¾in zinc plated steel (53g weight).
- ☐ Two (2) ⅜-16 x ¾in zinc-plated full-thread steel screws.
- ☐ Two (2) ¼-20 zinc-plated steel all-thread 5+in long. (Or bronze rod to be tapped.)
- ☐ A half dozen+ (6+) ¼-20 nuts & combination lock-nuts.
- ☐ Four (4) each 1in OD x ¼ ID chrome-plated flat washers.
- ☐ Four (4) 1in OD x ¼ ID soft rubber washers ⅛in thick.

---

[69] Note that the resulting prototype is not pretty. So when demonstrating in front of your audiophile friends, you are well-advised to *have them HEAR the arm BEFORE you allow them to SEE it.*

- ☐ One (1) ¼-20 stainless steel wing nut.
- ☐ Two (2) #6-32 long screws, 2in length.
- ☐ Six+ (6+) ea. #6-32 nuts, flat washers, nuts, & lock nuts.
- ☐ Several #6 x ½in Phillips steel tapping screw. (Select for polishing to a point for the pivot; keeps a spare for replacing point in future.)
- ☐ Two (2) ea. #4-40 screws ⅜in long & #4-40 lock-nuts; #4 flat washers.
- ☐ Two (2) 5/16in and one (1) ¾in self-stick felt pads.
- ☐ One (1) droplet of grease (prefer white with Teflon).
- ☐ 5/16 x 2in strap snipped from ⅛in aluminum flat stock.
- ☐ Salvage 1¼in long piece ½in ID rubber garden, air hose, clear PVC tubing, or ¾in OD neoprene power cable jacket). *Slit its length straight down one side.*
- ☐ Optional wood 1x4+ arm-board, cut to long dimension of turntable base, plus attachment hardware as needed.
- ☐ Pickup\stylus to work with variable-mass arm. *Tested over the range of 1¼~5g VTF with Stanton 500 & 680 with 10~25CU styli.*
- ☐ *Salvage an analog mouse cable 4-conductor+foil shield (very flexible stuff).
- ☐ Four (4) push-on cartridge connector "flea clips." *Do not solder to cartridge pins.*
- ☐ Two (2) #6 tinned spade lugs.

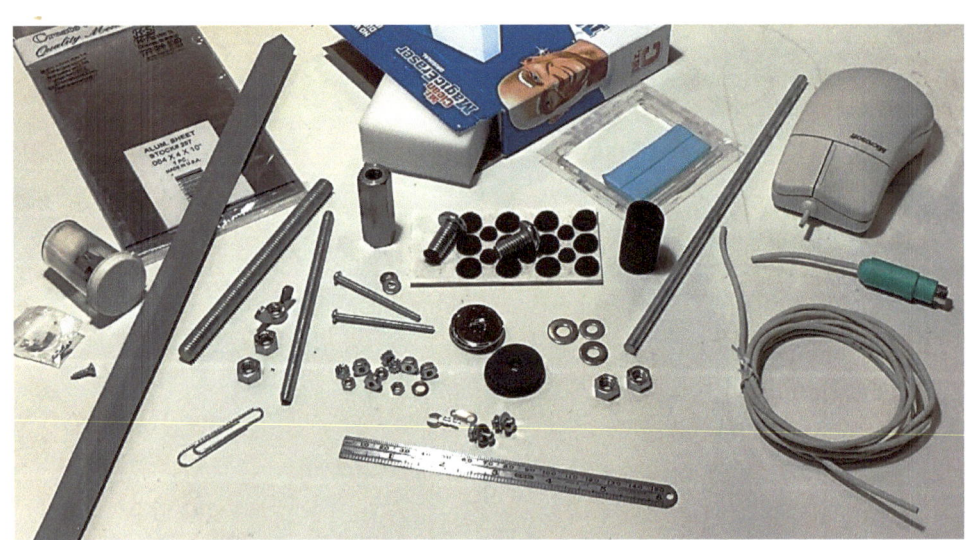

Parts for 12in tonearm are ordinary hardware. Clockwise from left: #5 (or #6) steel pivot screw, pickup flea clips, cartridge, ⅛in aluminum for finger grip, ½x½ in aluminum angle, ⅜ & ¼in all-thread, ⅜-16 coupler ⅝x1¾in, MagicEraser, ⅜-16 round-head screws & felt stick-ons, 1¼in neoprene cable jacket (or garden\air hose), BluTack, ¼in all-thread for arm-rest, what's left of the mouse and its harvested cable, 6in scale, 1in rubber & stainless washers, #6-32 x 2in screws & lock-nuts, and jumbo paper clip.

- ☐ One (1) heaping tablespoon of BluTack putty.
- ☐ A ~1in (~25.4mm) square of MagicEraser. (Whaa?)
- ☐ One (1) steel jumbo paper clip. (Whaaa?!!)

Now on a cleared, well-lighted workspace, organize parts, tools, reading glasses or magnifiers, bottled water, Ibuprofen, etc. Refer to the accompanying photos as you go. Mistakes won't cost much, so just settle in for a few hours of *hands-on fun!*

3. Fabricate wand sub-assembly [70]

This is the first of four groups §3~6 of construction steps:

- ☐ *Measure 1⅛in from one end of the aluminum angle and flatten in a vise. Twist flatted end to be 45° to each remaining flange.
- ☐ *Under the other end of the wand, mark from end ¾in, 3in, & 4in. Double check innermost mark is 12in to flatted end.
- ☐ *Clamp wand horizontally in the drill press vise, with the open "V" shape facing up, blocked with 45-45-90° wood chocks.
- ☐ *Drill the innermost marked hole at 4in using a 7/64in bit.
- ☐ *Drill two holes at ¾ & 3in from the end with a 9/64 bit.

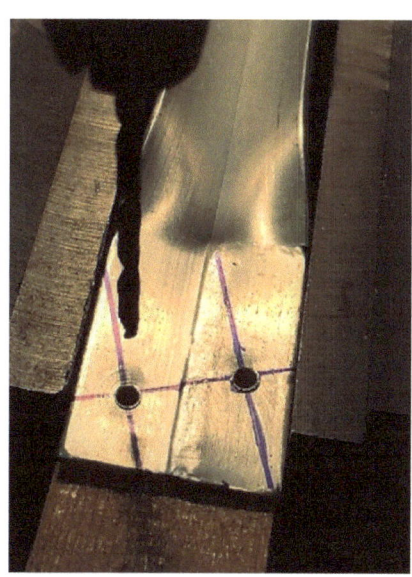 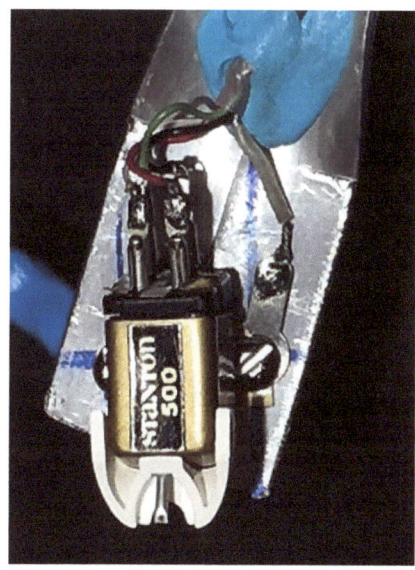

L: Underneath arm, intersecting lines mark 7/64in pickup mounting holes, with offset angle 18°. "Walking" of drill is negligible if it affects holes identically to preserve geometry. **R:** pickup installed with ground conductor is bonded beneath a cartridge mounting screw (do not insulate cartridge from the arm, but remove any ground strap). Blue-tac supports wiring and damps arm-stylus resonance.

---

[70] This paper avoids complicated work, harvesting hard-drive bearings, etc. As instructions might change, see - www.filmaker.com/papers/UPDATES_RMiller-Better Sound of the Phonograph.pdf .

A transcription tonearm – how come; how-to

- ☐ *Use a protractor, and refer to the photos. On underside of wand's flatted end, mark a line from the end corner on the spindle side at precisely 18° angle to wand length. *Sets the critical offset angle. Double-check angle from other side of the wand.*

- ☐ *Mark a point from the end corner precisely 1/8in more than the tip-to-ears dimension of the cartridge you intend to use. [71]

- ☐ *Mark a second line precisely perpendicular to first line.

- ☐ *Mark a point on the perpendicular line precisely 1/2in from the intersection of the two lines.

- ☐ *Mark a third line perpendicular to the second. *Confirm that it is at precisely 18° to wand's length, or REDO last 5 steps.*

- ☐ *Double check that the first and third lines above are precisely 1/2in apart (parallel) everywhere along the lines.

- ☐ *Center-punch the intersections, and drill 7/64in holes.

- ☐ *Carefully elongate both holes fore & aft along the 18° lines to make 7/64in slots ~3/8in in length. (The final positioning of the pickup will occur during alignment.)

Caution: especially using power tools, please wear safety glasses. Remove burrs with file & emery cloth after cutting\drilling. Clean with 30psi forced air, wash part & your hands with mild soap & water, towel dry part, and thoroughly air-dry with forced air.

- ☐ Only snug two (2) #4-40 screws & nuts from the underside (to mount pickup).

 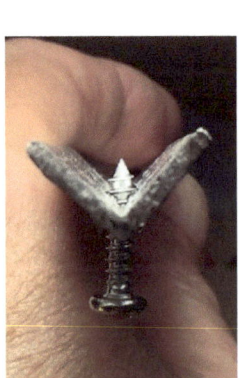 

L: Pivot screw hole 7/64in from underside of wand. 45\45\90° wood blocks in vise secure the angle. M: Polish uni-pivot screw in wand before washing debris. R: Drilled dimple atop pivot post prior to beveling outside. Arm is self-righting if uni-pivot screw and dimple in post are perpendicular & polished.

---

[71] For a Stanton 500 (Pickering V15) or 680\681 (XV15), this is 9/16in.

- [ ] *Its endmost threads dabbed with the tiniest drop of oil, partially thread #6 x ½in tapping screw from *underside of the wand* precisely perpendicular through the innermost hole. When threading just starts to go easily, hole is sufficiently tapped; remove the "pivot" screw. Do not wipe away oil on screw; wipe from wand.

- [ ] *Spinning in the drill press, polish smooth a cone and point of the #6 pivot screw (still with a bit of oil at the tip) using a fine file & emery cloth. Clean the screw using Dawn dish soap & water, and thoroughly dry.

- [ ] Partially rethread pointy pivot screw from *top of wand* until it protrudes underneath 3/16in. If the screw develops wobbly play, apply a drop of super glue or epoxy.

- [ ] Install from top two (2) #6-32 x 2in screws in wand holes at ¾ and 3in.

- [ ] Underside on each 2in screw, spin up a #6-32 flat and lock-nut. Only finger-tighten for now to allow for aligning later.

- [ ] *Mark & center punch center point on ⅜-16 coupler side.

- [ ] *Drill coupler through other side using a 5/16in bit.

- [ ] *Tap ⅜-16 coupler, hereafter called "counterweight."

- [ ] Thoroughly wash debris, and blow-dry counterweight.

- [ ] Clamp in vise between blocks ⅜-16 x 2¾in all-thread.

- [ ] *Mark dead center ¼in from each end straddling threads.

- [ ] *Drill through holes using 9/64in bit. Wash debris & dry.

- [ ] Spin counterweight about half-way onto ⅜-16 all-thread.

- [ ] Spin up ¾in on 2in screws #6-32 nuts + lock washers.

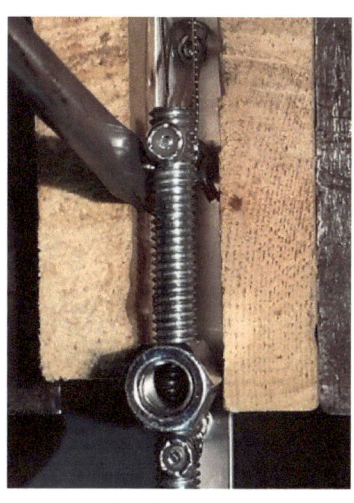 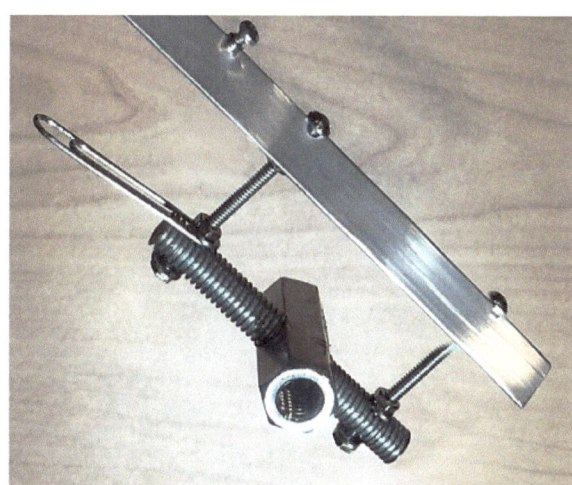

L – If counterweight all-thread is not aligned with wand, use a screwdriver to pry parallel. Counterweight ranges 1~5+g vertical tracking force (VTF). R – Wand from pivot back, featuring the revolutionary "anti-roll & dive control" paper clip.

- ☐ Add a jumbo paper clip at one end, double-loop side out. *You'll see why, just wait!*
- ☐ Slide all-thread with counterweight up arm's 2in screws.
- ☐ Add two (2) lock-nuts. Slide all-thread down 2in screws and adjust lock-nuts flush with ends of 2in screws.
- ☐ *Align counterweight sub-assembly inline with wand. Tighten six (6) nuts, aligning paper clip on-axis with all-thread, clamped at the end of the clip's single-loop.
- ☐ Stick two (2) 5/16in felt pads on the ends of ⅜-16 x ¾in all-threaded screws, and trim felt edge inside the screw diameter.
- ☐ Thread ⅜-16 x ¾in felted screws in counterweight ends.
- ☐ Test that felted screws arrest counterweight spinning.

This completes the wand sub-assembly; set it aside. *Congratulations!* You qualify to finish the project, as no further steps are more difficult. Let's take a break *and solder!*

4. Wiring (up to preamp connectors, which are user's choice)

- ☐ Power 30W soldering pencil; allow to reach temperature (test melt a bit of solder).
- ☐ Meanwhile at each end of mouse cable, strip away 1½in of jacket. Strip away foil shield 1½in. Unravel four (4) twisted wires and strip conductor insulation 3/16in.
- ☐ Clean soldering tip on damp sponge. "Tin" by applying thinly #22 flux-core electrical solder (prefer 63\37 eutectic).
- ☐ *Twist the individual conductors of five (5) wire ends. "Tin" thinly with solder.
- ☐ * *If heat-shrink tubing is available, slip ¼in of ⅛in tubing well back along each signal wire (4). Slip a ⅝in length of ¼in shrink tube over four (4) inner wires and well back from the end of the cable jacket.*
- ☐ * Form a hook in each tinned end, pass through eyelet of each pickup connector flea clip (4), gently crimp, & solder. *Slide ¼in of ⅛in tubing over soldered wire of each flea clips (4), and heat to shrink. Slide ⅝in length of ¼in shrink tube just back from the end of the cable jacket, and heat.*
- ☐ *Similarly at preamp end, solder signal wires (4) to chosen connectors observing the order that follows for the cartridge signal pins: left pair opposite, also right pair, no pairs side-by-side.[72]. Solder shield to a #6 spade. *Affix shrink tube as above (5).*
- ☐ *With wand down-side-up, arrange cable with a bit of excess at the cartridge end. Secure cable smushing a dollop of Blue-tac at ~four intervals along underside of wand. The last dollop should be ~1in in front of pivot.

---

[72] This is quasi-"star-quad" configuration to minimize crosstalk & hum pickup. (Standard colors: L+ white, L– blue or black, R+ red, R– green\yellow.)

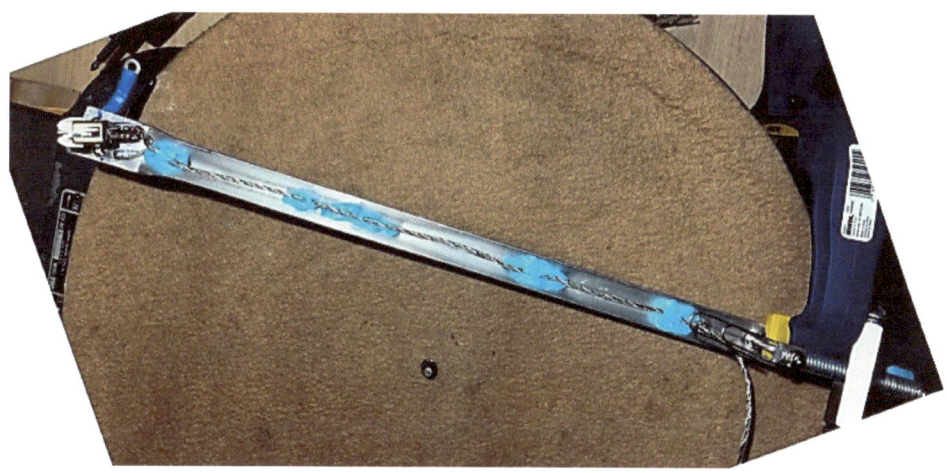

Underside of finished wand. Shown for high RFI environments is optional 100% braid-shielded 4-conductor phono cable, with grounding eyelet under one cartridge mount screw head. Secure cable at random points with BluTack, ending ~1in forward of the pivot screw. **Shield should not make any contact along wand or pivot post.** Putty may be added, subtracted, or moved to damp resonance. Lower the resonance frequency by moving\adding tack toward pickup, or raise it away from pickup.

- ☐ From the last dollop and the pivot toward the free end of the cable, remove jacket and shield foil ~4in. Splay five (5) exposed wires, including bare shield conductor for maximum flexibility while preserving loosely twisted pairs (cf. image **p134**).

- ☐ Under the #4-40 screw heads underneath the head of the wand, install the pickup cartridge, with stylus removed for its safety. If moving iron type (MI), tape over mounting tube to prevent magnetic particles entering. Press on the four conductor flea clips to correspond with the L+, L-, R+, and R- wires at the preamp end.

5. *Fabricate finger lift; armrest & clamp sub-assembly*

- ☐ *Drill two (2) ⅛in holes ½in apart at one end of the 3/16 x 2in strap (finger-lift).
- ☐ *Opposite, form a finger diameter inverted "U" (¾in squat "Ω").
- ☐ Install atop wand's flat end under the #4-40 nuts for holding the cartridge.
- ☐ *Inside the rubber hose\tube pried open as slit, drill a ¼in hole ⅛in off-center.
- ☐ Screw from outside the tube the 5in ¼-20 all-thread to protrude ¼in inside.
- ☐ Fasten inside tube with ½in stainless flat washer and ¼-20 nut. *Adjust until end of all-thread is flush with nut,* and the tube's slit is forced slightly open.
- ☐ Fasten outside tube (its bottom) with larger stainless washer and ¼in lock-nut. Tighten until tube slit opens ~⅜in.
- ☐ Apply ¾in felt stick-on to all-thread & nut inside tube.
- ☐ Loosely add ¼in nut, 1in stainless washer, two (2) 1in rubber washers, 2nd 1in stainless washer, & lock-nut (see photo).

A transcription tonearm – how come; how-to

6. Tonearm installation on base, alignment, & testing

Next, prepare the turntable base, or if cramped an external arm-board alongside, minimum size ¾ x 3½+in. If the only arm, locate it along right side of turntable; if a 2nd arm, locate at back facing left. On base or arm-board, mount both pivot post & armrest:

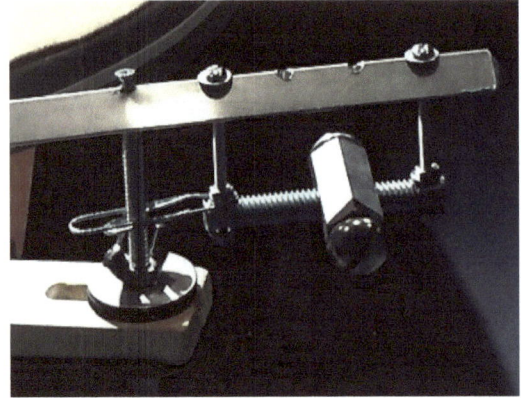 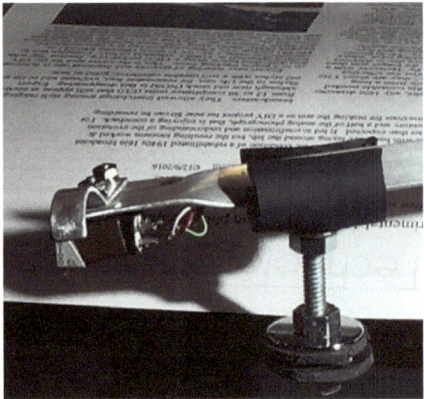

L: "Roll & dive control paper clip" is snug along sides of pivot post without snagging on threads. For mounting the pivot post in the turntable base (plinth) or added arm board, drill two ⅜in holes and using a coping saw elongate a ⅜in slot between them as instructed below. **R:** ~2in from the arm's end, drill a ⅜in hole in the base for mounting the armrest. Arm is returned to the armrest one-handed by prying open the longer side of the tube clamp with your right index finger, and the shorter side by your right pinky.

Fabricate the tonearm pivot post; locate it on the base

- [ ] *Near the desired position for tonearm pivot post, mark a short arc of radius 11¼in from the spindle, and a second arc of 12¾ in.
- [ ] *From the desired position of the pivot post, draw a line intersecting the two arcs at one end to the approximate midpoint between the spindle and the imagined armrest.
- [ ] *At each of the two (2) arc intersections, drill a ⅜in hole (2).
- [ ] *Using a sabre, scrolling, or coping saw, complete a ⅜in slot between holes at the arc intersects. Finish for easily sliding the pivot post assembly, fabricated next.
- [ ] *Precisely punch dead center at one end of the 5 in ¼-20 all-thread (or 3/8 bronze rod) for a pivot post. Clamp vertically through a drill press vice using wood blocks at each side near one end. [With access to a metal lathe, use it for next two steps.]
- [ ] *With a ⅛in bit, drill a dimple atop the post (see R photo p108). Slowly continue to a depth of ⅛in, pausing at bottom ~10s maintaining light pressure to polish bottom.
- [ ] *With post spinning in the drill press, grind an outside bevel using a metal file and polishing with Emery cloth, form steep conical sides to just outside the periphery of the dimple. Wash debris with soap & water, and thoroughly dry the post.
- [ ] Thread a ¼-20 wing nut (wings up) ~1½in up post from bottom. Add 1¼in chrome washer and rubber washer. (*If using a 3/8 bronze rod, tap ⅜-16 its bottom up 2in.)
- [ ] *Adjust wing nut up\down to raise or lower top of pivot post, so wand will be horizontal *with cartridge resting on a disk. (Sets stylus Vertical\Rake Angle.)*

- ☐ Insert the pivot post through the base slot, underneath snugging with a second rubber washer, chrome washer & lock-nut.
- ☐ On base underside, hand-tighten a second ¼-20 nut against the first. Using two box wrenches, lock the nuts together. *Hereafter during alignment, use the topside wing-nut to slide the pivot post in the base slot to adjust stylus overhang.*

Locate arm rest on base; install wand

- ☐ To locate armrest, mark base or armboard 9 to 9½in ahead of the pivot post, comfortably away from the turntable platter (or disk as large as 16in if desired).
- ☐ Drill ¼in hole, install pre-assembled armrest, with rubber washers either side of base\armboard, aligning slit toward pivot.
- ☐ Place a tiny droplet of Teflon grease in the pivot post dimple. *Mop up any spillage.*
- ☐ *Preparing to install wand, adjust using pliers the inner loop of paper clip "roll & dive control device" *just snugly around pivot post, but free of snagging any threads.*

Exercise caution anytime you move the arm wand, as the pivot point can jump off the pivot post and cause damage. If ever at any risk, remove stylus from cartridge. <u>Anytime tonearm is not playing a record, return it to armrest</u> and secure in clamp.

- ☐ *Slide wand assembly on pivot post. Adjust pivot screw just enough for no wobble.
- ☐ Re-test & readjust paper clip inner loop for *no rolling* of the arm about its axis when at record height. *Test arm swing for no resistance, horizontally or vertically* until it reaches the dive limit of the paper clip.
- ☐ *Examine from in front whether cartridge is precisely perpendicular to disk surface. To adjust, *forcibly twist arm near pivot to bend "roll & dive control paper clip."*
- ☐ *Install the stylus, exercising care in the steps that follow.
- ☐ *Check pivot post locknuts under base and wing nut on top so tonearm is parallel with a disk. Tighten lock nuts, then snug wing nut, slightly compressing rubber washers. *Recheck arm is parallel to disk!* Remove arm to clamp in armrest.
- ☐ With your target vertical tracking force (VTF) in mind, rough-position the arm's spinning counterweight. (You will fine-adjust it when you later optimize the arm\stylus resonance.)
- ☐ Lay a 1in square of MagicEraser on a stylus pressure gauge or kitchen scale aside the platter on the base (or use a 2.5g penny balanced on a cardboard "see-saw"). *Set the stylus on the MagicEraser and read current vertical tracking force (VTF). [73]

Shim under gauge so MagicEraser is flush with record, or VTF will read in error!

- ☐ Move arm to armrest. Loosen counterweight locking screws. Spin counterweight an estimated distance (range 1~5+g) *outward to lighten, inward for heavier VTF.*

---

[73] MagicEraser cleans styli by poking it several times.

- ☐ *Repeat measure until VTF is within stylus manufacturer's spec. (Range for this arm, allowing compensation adding cartridge weight, is 1¼ ~3g.) *Snug end screws.*

- ☐ *Print p1 http://www.filmaker.com/papers/UPDATES_RMiller-Better Sound of the Phonograph.pdf 100% of size (Scaling = "none"). On paper if it's precisely 1.50in square cut it out. On the underside apply adhesive tape to cover the diamond shape.

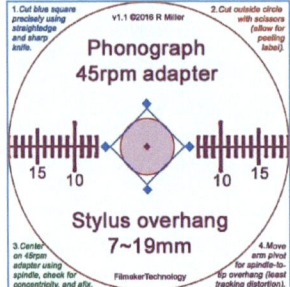 

Tonearm alignment achieves tracking distortion of 0.47% maximum. Aim the Overhang scale toward the arm pivot, parallel to the DIY arm wand, not the side of the cartridge. Slide the pivot post until the stylus tip rests on the 13mm mark opposite the post.

- ☐ *Using a sharp (x-acto) knife, slice along blue lines, and remove diamond shape. (Use scale above for 7~19mm overhang distance of the stylus past the spindle.)

- ☐ *Center the gauge on a flat 45rpm adapter. (Consider permanently gluing it?) Rest the stylus gently on the overhang gauge, with graticule in line with and opposite the pivot. Read and note the "current overhang" is ____ mm. *Return arm to rest.*

- ☐ *Loosen pivot post wing nut; taking care of the stylus, move the arm by its post in its slot until stylus tip rests on 13mm line. *With post centered in slot so it makes no contact with base, snug wing nut.* With arm in rest, tighten wing nut. *Check 13mm.*

- ☐ *Arm might skate across record due to torque from wiring. Allow arm to descend and float above base. Ensure wiring is not *biasing* arm inward; slightly outward is desirable. Wrap wires in a loose coil counterclockwise around post to provide anti-skating bias. Tighten coil radius to decrease, or loosen to increase bias until the cantilever is not tugged in or out from the spindle when playing a running groove. Secure (wire-tie) cable beyond the coil. *Recheck for no skating.*

  Skating redirects ~13% of VTF, so increase static VTF by that amount.

- ☐ Preamps vary in connections. Consumer audio uses two "RCA" unbalanced phono plugs plus a ground wire. Professional installation preserves the pickup's balanced wiring + ground with a 5-terminal block (four (4) signal wires plus ground).

- ☐ Connect five (5) wires to preamp inputs and ground lug. (Each RCA plug has both + and – connections.) Or clamp 2 twisted pairs + ground(s) under screw terminals.

- ☐ If preamp has selectable capacitive load, select its input C_{preamp} per the cartridge spec., minus cable capacitance. *Results in best frequency & phase responses.* [74]

---

[74] Details are in the previous "Tuning…" chapter, with maker instructions for a cost-effective phono stage.

- ☐ Place a sacrificial record and note how the arm tracks the groove. Is the stylus pulled to one side of its grip (skating)? Does needle jump when mildly thumping the base (resonance)? If so, double-check that anti-skating bias is not forcing the cantilever off-center while playing a groove.

- ☐ If the preamp has individual channel gain controls, play a monophonic record and adjust channel gains for equal output levels. *Balance results in best soundstage for stereophonic records, and lowest vertical distortion for monophonic disks.* [75]

- ☐ Again play sacrificial record. Does needle jump when mildly thumping base? If so, increase VTF (not above mfgr's spec.). *Also see resonance adjustment below.*

This completes construction of the Steampunk 12in Transcription Tonearm. [If as recommended you are pre-reading this prior to beginning construction, finish reading this and the next paragraph, then return to **p110 §2** to begin work.]

True, for $600~2500 and up (the sky's the limit!) you can *buy* a better arm than you just made. Or can you? Assembling dozens of kits as a teen (and designing & building many more custom devices since), I have found much greater value than if I had had no hand in making them. And experienced much greater satisfaction lasting years compared to only hours of pleasure unwrapping store-bought stuff. Even if this is your first such foray, if you have (or can borrow?) the tools, or recruit the person with the tools to help?! You'll likely find this project only mildly to moderately challenging. And soon, to *Enjoy!*

7. Fine tuning – optimizing arm resonance, skating compensation

If you were successful following the instructions above, your experimental 12in transcription tonearm will likely work, and sound amazingly good, as did my first (and two more). But a tonearm is the closest partner of the turntable's critical *energy transducer*, the pickup cartridge, and this pairing is attended by many interdependent variables. Before playing other than a sacrificial record – and remembering to return the arm to rest whenever not playing a record – fine-tune these variables, as instructed next in *italics*...

1. Check that the cartridge (not a twisted stylus!) is perpendicular to the record surface. Note that many cartridges do not have parallel sides. *Twist the arm wand about the pivot point to bend the paper clip until the cantilever and its tip are precisely plumb.*

2. Check that there is no rolling – no visible or feelable wobble. *Adjust critically by gently bending the inner loop of the paper clip to barely touch the all-thread, yet that it doesn't impede in the slightest the arm's movement, horizontally or vertically.*

3. Check from the side that the arm (not the cantilever) are parallel with a record on the spinner's platter. *Loosen and adjust the pivot post's locknuts under the base, and wingnut above. Retighten all, noting overhang has likely changed.*

4. Check the tracking pressure *with the reading surface of the gauge at record height.* Set VTF for the stylus in use (see cartridge specifications – typically 1¼~2g for higher compliance styli 18~25CU, 2~3g for medium 10~18CU. If the counterweight cannot reach the lightest desired VTF, add a ⅜-16 nut etc. at the back of the ⅜-16 all-thread.[76]

---

[75] Ibid. The "Tuning..." paper also describes a DIY $35 preamp with both C-load and gain balancing controls.

[76] For the low mass 12in arm as built, use styli specified 18~25CU tracking 2~1¼g. Add mass for styli 10~16CU and VTF ranging 2¾~5g. Note that Expert Stylus UK recommends VTF less than 3g.

5. Check overhang is 13.0mm. Viewing from above straight down, check that the stylus cantilever (not the body of the cartridge behind it) now remains tangent across a 12in LP. (This verifies 13mm overhang & 18° offset angle.) *Adjust by loosening the pivot post wing-nut and moving post in slot in base, relocking to have no contact with base.* ***Best is a mirror-surface alignment protractor to view a stylus from underneath.***

6. Using a protractor, align (aim) the cantilever by the cartridge mounting screws in their slotted holes. ***Best to use a mirror-surface protractor to view stylus from underneath.***

7. Check that wiring is not biasing the tonearm inwardly; slightly outwardly is preferable. *Adjust the coil of five (5) conductors as in the "Skating" paragraphs* ***below****, and the illustration on* ***p62***. Wire-tie gently only on the side of the coil farthest from the pivot.

8. A remaining critical adjustment is the *resonance* of the combined tonearm & stylus. If not optimized, resonance can produce feedback, boomy low bass, accentuate rumble, and cause the stylus to lose contact with, or jump the groove. *Optimize resonance to 10~20Hz (some advise 8~12Hz) by varying arm "mass" (moment of inertia), cf.* ***p136***.

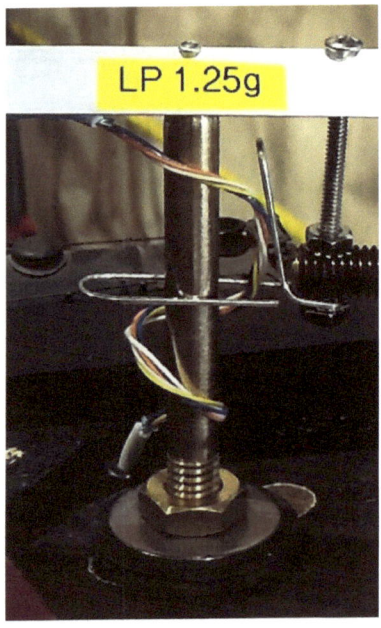

4-conductor+bare ground harvested from computer mouse cable. Free of jacket and foil shield, it is quite flexible. Coil 1½~2 times CCW around the post to vary anti-skating torque by loosening or tightening the coil radius. (Prototype #3 shows a ⅜in bronze post for its bearing properties.)

Setting anti-skating bias. As on **p62**, skating is caused by variable vinyl coefficient of friction. It diverts a share of VTF to a force on the inner groove (L) wall, and less on the outer (R). Skating increases wear of the inside groove wall and stylus. And can increase distortion on the right channel from loss of groove contact. A 12in arm's skating force is ~62% of that for an 8in arm (73% of a 9in arm), and that much less sensitive to differing disk friction. Some hobbyists err by increasing VTF, worsening uneven wear. Best is to apply anti-skating "bias" force. But not too high, or skating effects will switch to the opposite side. Skating redirects VTF, so increase static VTF by ~13% to prevent distortion.

The questionably elegant solution is in keeping with the steampunk arm. Just as with the spiral spring provided in factory-made arms, it involves wrapping the delicate tonearm wiring counterclockwise 1½~2 turns around the pivot post, shown **opposite**, then loosening or tightening the coil to adjust its torque. Adjust anti-skating by observation, one visual, one audible. Easier is visual. If while playing the disk the cantilever is pulled off-center toward the outside, the culprit is skating tugging the arm toward the spindle. If the stylus is off-center toward the spindle, bias is set too high. But it can change with the next disk!

Audibly, we become aware of distortion if it differs between channels; the channel with lower distortion serves as a reference. Select a disk with a high level of centered (or mono) content. Listen critically for any distortion, such as so-called "sibilance" (spitting sounds accompanying vocal sibilants). If with high levels of centered (tending mono) sounds you hear more incidents of distortion in the R channel than the L, then skating force is tugging the stylus intermittently away from contact with the R (outside) groove wall. Loosen the coil diameter for more torque. If distorted peaks are higher on the L, anti-skating is set too high, the stylus pulling away from the L (inner) groove wall. Tighten the wire coil for less torque. Anti-skating is about as good as it can be when no difference in distortion is heard.

In case you feel that this anti-skating is crude, consider that skating varies all over the place from disk to disk, depending mostly on not insignificant differences in the coefficient of friction of the disks you play. More *elegant* methods, using pulleys & weights or a dial and spring, are little better, except perhaps for looks, cf. **p62**. Variable disk to disk, the visual and audible methods above are most effective for changes in skating.

Now it's safe mechanically to play a good record. However, the sonic qualities of the cartridge chosen for the arm (frequency & phase response) will depend also on proper alignment, including optimizing resonance, as discussed. Also as mentioned for MI or MM pickups, a C-load mismatch can cause sound to be too dull, or too bright; bass weak, or too booming. Imbalanced L\R levels ruins the soundstage in stereo, and passes distortion in mono and vertical rumble in stereo. Solutions are in the appropriate prior chapters.

As work of this book concluded, I built second and third prototypes, following the instructions above. (I even disassembled and my grandsons rebuilt the arm.) It took 5hr taking my time, as should you, dear reader. A med-high compliance stylus of 25CU tracked at 1.0g. This is good performance for any 12in arm. The uni-pivot design has proven worthy, simplicity itself, and apparently requires no exotic materials or expense. Walter O. Stanton himself is the inventor, having applied for the Patent in 1957.

8. Concluding construction

Properly aligned, the experimental *pivoted* 12in tonearm has offset angle 18° and overhang 13mm, within <±¼° and <±¼mm (about as fine as you can set them). Very early straight arms (today used by scratch DJs) of 0° offset angle and 0mm overhang create(d) 5~10+% distortion, deemed acceptable until the high-fidelity era >1948. Then expectations improved an order of magnitude to <1%. [77] 0% distortion – no *coloration* – is the objective goal still for some, while for others the subjective preference is "euphonious distortion."

It is this *euphonious distortion* that many audiophiles champion online. Echoing the purple prose of magazine reviewers, they use undefined words like "air," "warmth," "soundstage," "liquid midrange," "black background," "3D" (for the "1½D" sound of two speakers 60° apart in only the horizontal plane, hardly a 3D sphere), etc. The only thing 3D about 2-channel stereo comes from the listening room wall, ceiling & floor reflections, not the recording. The objective of this book is to reduce distortion artifacts to a base level.

---

[77] Tangential-tracking turntable-tonearm units have zero tracking angle error, no distortion due to it.

For the Steampunk tonearm in the simulation below, dimensions are boxed **in yellow**, alignments **in green**, total harmonic distortion (THD) due to tracking error for various disk sizes & speeds is calculated in red. *Three maxima of 0.47% across a 12in stereo LP, and minima of 0% at two null-points, where the cantilever is aligned tangent to the groove.* A 9in tonearm can be aligned for LPs to 0.67% THD. For 16in ETs, even at their higher linear velocity, max is 1.25%, 1% for 7in 45s – inaudible over AM radio where ETs & 45s were employed. [78] At 78s' still higher linear speed, THD is 0.94% max. THD percentages imply likely comparable, but worse sounding intermodulation distortion (IMD).

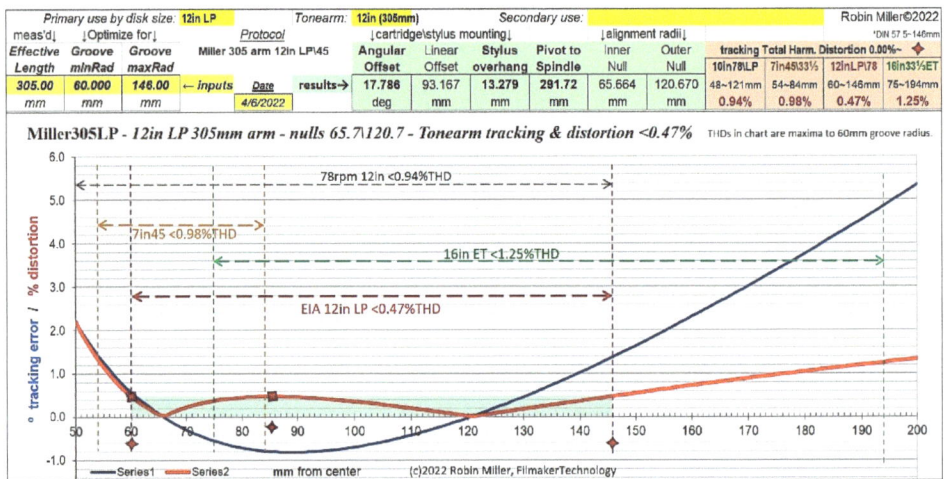

Lofgren-A\Baerwald alignment of the 12in (305mm) steampunk arm calculates an offset angle 17.8º, overhang 13.3mm. Distortion (red curve) for a 12in LP is 0.47% max. Adjusted for 78 & 45 speeds, THD is <1% max (red in red boxes). For 16in ETs, while outer groove shows highest angular error, high linear speed reduces max THD to 1.25%. Cf. more common 9in (225mm) tonearm data on p59.

Explored in the *Stylus close-up* chapter, there are other grooved media distortions and artifacts, most perceivable playing recordings of acoustic instruments and the unprocessed human voice in classical, jazz, folk, choral, and spoken word genre, etc. We unconsciously refer to our experience hearing these sources live – a *reference* in memory that is distortion-free. On the other hand, reproduction of electric instruments – heavy metal, etc. – offers little or no reference, as each overdriven guitar amp or produced effect sounds different.

For any tonearm, pivoted or not, tracking distortion may well be masked by the *tracing* distortion of poor performing styli, pinch-effect (2nd harmonic, out-of-phase in stereo), *poid* tracing (3rd harmonic), or prior wear of groove or tip. Artifacts are worse with a spherical stylus playing inner grooves, where groove speed is half that outside. For monophonic records, properly mixing L & R preamp signals cancels pinch-effect, therefore monophonic reproduction from a groove (mono or stereo) is inherently cleaner than stereo replay!

For the DIY arm's effective length of 12in (305mm), the alignment spreadsheet above calculates an offset angle of 18º and an overhang of 13mm (green calculation boxes). The arm's design accommodates all disk sizes from 7in 45s to16in ETs; it is optimized for the 12in LP at 33⅓rpm (red curve). This compares with 0.67% max for the well-regarded

---

[78] Broadcast and professional gear manufacturers once considered IM distortion of 10% "inaudible," until the hi-fi era! Today's components (except speakers) can achieve distortion as low as 0.01%.

SME 3009ii arm – see its similar spreadsheet on **p61**. Aligning for low tracking distortion is especially sensitive toward the inner groove region, where linear groove speed is lowest.

While these distortion products are *not recorded* in the groove, and so are avoidable by better replay, there are other distortions baked in during recording, mixing, mastering, and pressing processes. This is why some disks sound good, others not so. These might take us back to reproduction distortion of 10+% overall. Before loudspeakers add still more! [79]

Frequency & phase responses, heard as incorrect or correct tone color and smeared or crisp transients, are dictated by the user's combined pickup & preamplifier ("phono stage"). Since 1953, most records are mastered using the "inverse RIAA" characteristic, which must be precisely reversed in the preamp. [80] Prior to standardization, record labels used hundreds of combinations of Turnover, Rolloff, & Rumble filters, and kept them trade secrets. [81]

Users and reviewers too often overlook the need for properly loading a moving magnet (MM) or moving iron (MI) pickup. The resistive load is mostly standardized at 47kΩ. In parallel, the load capacitance is typically ~275pF (pico-Farad), and is the sum of *distributed lumped C* of the wiring and the *C* inside the preamp. As was shown on **p35**, a preamp plus cartridge can achieve excellent tone color & transients when flat ±¼dB 50~15,000Hz. [82]

The opposite of these admitted imperfect (but usually verifiable) objective measures, audio magazines have popularized subjective opinions, touting in vague unsubstantiated terms which coloration (distortion or altered frequency response) "sounds best," according to the self-appointed *golden-eared*. Advertisers' hype, and inverted price-value claims, should be taken with a grain of salt. Marketing research shows that listeners become perceptually habituated (conditioned), remaining fixed in taste until re-acclimated. Like one who has used bad grammar all his\her life thinking something said correctly sounds wrong. My dad taught me: "Better to *question the answer* than to answer the question." I trust much more the subjective opinions of evaluators who use unbiased double-blind methods, statistical analysis of correctly chosen subjects (not tired students or inexperienced auditioners), and who try to confirm any subjective opinions with the admittedly imperfect objective measurements that engineers prefer to rely on.

I hope you enjoy making and using the DIY 12in tonearm – there is great experience and satisfaction to be had. And many years to enjoy playing phonograph records better.

9. *Steampunk tonearms "resonate" far & wide*

Shortly after making the 1st DIY steampunk 12in tonearm and drafting this book, the author's 2nd and 3rd prototypes confirmed the instructions, alignment including the anti-skating bias, and performance. The arm is low in "mass" permitting use with moderately high compliance cartridges at lightweight VTF, tested without skipping down to 1g!

Measure arm-stylus resonance: For any turntable, first attend to plinth\base resonance. With the preamp and amplifier ON but the turntable still, firmly wrap on its base with a wood\rubber handle screwdriver. This is called an *impulse*. Listening close to a quiet

---

[79] Also listening room acoustics, speaker crosstalk comb-filtering, etc. Replicating live hearing requires full-sphere 3D reproduction, e.g. the author's 10 speaker HSD-3D – see www.filmaker.com/papers.htm

[80] EIA standard added, and later repealed an infrasonic filter <20Hz (for replay only, not mastered).

[81] For RIAA, three filters are specified with –3dB ("f₃") points of: turntable "Rumble shelf" filtering is below 50Hz, "Turnover" is 500Hz, and "Roll-off" is above ~2kHz, down 13.7dB at 10kHz.

[82] Cf. the DIY preamp. I don't plan (yet) to make a Cadillac transmission flywheel into a turntable!

speaker. Preferable is a dull thud. If a decidedly gong- or (worse) bell-like ringing, and (worse) if it lasts more than a second, secure to your base some acoustically inert stuff the mass of lead or concrete to dampen it. Now attend to the critical arm-cartridge resonance.

To prepare to measure and optimized resonance, we will employ the *impulse response* method. Download & install *Audacity* (freeware audio recording\editing app). Familiarize yourself with the wave *Spectrum (*frequency response) viewer, noting how to observe VLF between 3~100Hz. Or borrow a digital oscilloscope (DSO), or a storage scope of yore. [83]

Plant the stylus in a silent groove (lead-in or run-out). After several decreasingly gentle taps toward a clean response, the display shows LF "ringing' (if not attenuated by preamp rumble filtering). If this *impulse response* measures >20Hz, often an issue as elastomers age, lower the frequency by increasing the tonearm's mass (or by using a higher CU stylus). Gently lift the arm wand free, and turn it down-side-up. Add BluTack close to the cart, or a cartridge weight, re-install the arm (or another's headshell), reset the VTF, then remeasure. Repeat as necessary. If less than 10Hz, decrease tonearm mass, or use a lower CU stylus.

In the DSO's display along the horizontal axis, the steampunk arm's lateral resonance is "15.87" Hz (bottom R in display), reduced from an initial 19Hz that approached the high limit after adding washers of decreasing sizes under the cartridge mounting nuts atop the wand, plus readjusting VTF. (The arm project is intentionally higher within the 10~20Hz range because it is easier to add mass than to subtract it.) And ideally *damp the response* within the fewest cycles. A typically octave higher vertical resonance described on **p64** is less important when a disk is monauralized vertical (no modulation) <150Hz in mastering.

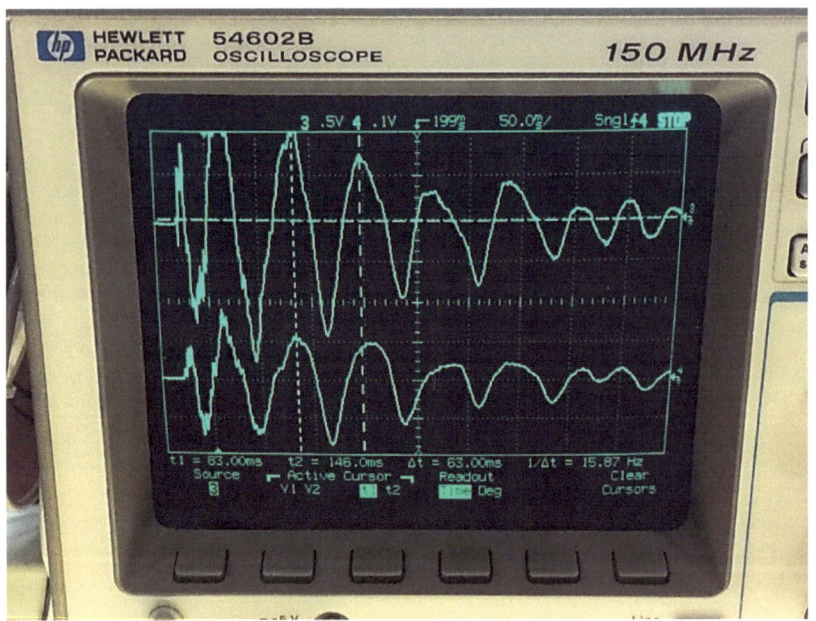

The final touch for any tonearm is optimizing arm-stylus-platter *resonance*. Use a free app's "spectrum analyzer" (or Digital Sampling Oscilloscope (DSO) above) to adjust, here an acceptable 15.87Hz.

As styli vary in compliance (CU), changing styli alters resonance. Whenever changing a stylus of cartridge, it is prudent to re-measure and, if need be, readjust arm mass. (For critical archiving, I don't trust "marketing" specs, and compliance changes with elastomer

---

[83] Useful with a spectrum analyzer function to set balance, channel-to-channel phase, rumble, THD, IMD...

aging.) Tonearm mass may be added, subtracted, or moved. Underscoring that resonance can vary with different styli, even those specified as direct substitutes, the new and new-old-stock I've measured ranges 12~16Hz. But in the same arm, a new N500-6 by the modern incarnation of Stanton resonated at 22Hz, mark of an inferior product that, playing the low bass of acoustic music, would be troublesome without increasing arm mass.[84]

And the author has received raves from happy listeners, who want to know what they can make next. Making can be more fun than shopping, as the satisfaction lasts a lot longer.

"Hooked" on sound much better than mp3, a reader posts to FaceBook friends his Steampunk 305 tonearm on his custom base with a used Stanton 680 cartridge and NOS 0.3mil elliptical. The vertical counterweight orientation is correct for reducing vertical resonance using a stereo pickup.

A $35 Goodwill find mounted on deeper bespoke base to accommodate the 12in steampunk arm tracking 2g. The direct-drive turntable's factory 8½in arm is still usable for a second disk format with another cartridge.

---

[84] Due to too low compliance (or unspecified) of more recent cantilevers.

Better sound from your entire audio system

You have arrived at the end of the book, and the beginning of optimal quality and increased enjoyment from vinyl LPs, 45s, and shellac or vinyl 78rpm records. Access to nearly a century and a half of history in music, broadcasts, and speech (drama, comedy, etc.). *Only the printing press exceeds the phonograph as an archival medium of human experience.*

Many publications cover the content available on phonograph records; the equipment in magazines and online. This book is about the science of *phonography*. The result of 5 years of initial research, then 5 more of 60+ years total professional experience in audio. For better implementation by hobbyists, DJs, broadcasters, archivists, musicians, historians, and the techno-curious. Have no qualms if you don't fully read, or do all that in this book – there's much to choose from, to come back to, to make a difference.

Not so long ago the phonograph was considered obsolete. Now it is making a comeback due to its historic archive. To new interest in collecting and listening to different genre: classical, American Songbook, jazz, and other acoustic music for which we have a remembered reference for verisimilitude. To clearer sound sans excessive audio processing. To the fun of "user serviceable parts inside." And gently setting a rock in a rut.

Of course the phonograph \ gramophone \ turntable \ record-player is only one component of an audio system, and new publications deal with sound sources, the evolution in digital power amplifiers, and especially loudspeakers & listening acoustics. The music in Gramophone magazine. And the many sources of expertly curated (DJ'd) music online and on-air.[85]

Spinners are robust, and cartridges do not wear out. It's the replay stylus that requires any maintenance – and that is most critical to the sound of the phonograph. Disks that require cleaning. The recent increasing trend in demand buying disks and the gear to play them will mean new demand will be met for styli of the best form: high-performing, low-wearing line-contact profile. But user knowledge is needed to choose, and to check purveyors making and selling them. Knowledge that is the reason for this book.

Whatever *your* level of interest, our best wishes for your endeavors in pursuit of better sound quality of grooved media. The nature of science is that your author may again find new knowledge, so the "Updates & extras" at www.filmaker.com/papers.htm will be revised as needed for the reader to check periodically. *Many thanks for reading.*

---

[85] Who will assume the mantle of America's favorite classical DJ, now retired after 50+yr at KUSC, Jim Svejda?

UPDATES & extras (hyperlinked content) – www.filmaker.com/papers.htm

Revised as needed, the free downloadable "UPDATES & extras" for this book contains many more illustrations and discussion (the last Update prior to this printing is 50+ pages). It expand on topics related to and beyond the phonograph to the total audio reproduction chain generally: from alignment tools to magnetic tape for production and disk mastering to digital mixes & intermediates for new vinyl releases\reissues, and from amplifier power and speaker wire and other "interconnects." Recent *UPDATES & extras* chapter titles are:

- Alignment tools [cutouts for an overhang badge for 45 adapter and speed strobe];
- The aftermarket for generic replacement styli; Pfanstiehl nomenclature;
- What about ceramic & crystal phono pickups?;
- Moving coil (MC) cartridges – darlings of "high-end" stores and online groups;
- A brief history of sound reproduction;
- The very first stereo records 1952, before stereo in a single groove;
- When The Golden Age of Hi-Fi ended – and will there be a 2nd?;
- Reference-quality sound, like a car, starts in neutral;
- Don't forget tonearm, stylus, & cantilever alignment;
- What is behind claims of "3D sound?";
- What about Richard Murdey's VSPS?;
- Electro-magnetic sound recording – the other analog medium;
- Related reading – too short a list of seminal and lesser known works on sound;
- How tomorrow might stereo evolve?;
- New released records & reissues (from original tape OR new digital);
- What to do if capacitive loading is too high? Too low?? Compensating with EQ???;
- Where to find the best new and new-old-stock (NOS) phono stuff for reasonable $?;
- Subjectivists v. objectivists – ne'er the twain shall meet?;
- Besides the record player, what about the other links in the audio chain?;
- Subwoofers, and 5.1 surround;
- Choice of loudspeakers, and the amplifier power they need;
- Interconnects (cables) and loudspeaker wiring;
- The future of audio reproduction: WaveField Synthesis? BACCH? Ambiophonics?

Grammy-winning mastering engineer
Clair Dwight Krepps

Near 100, C. D. Krepps shows an input strip from his 8-channel mixer.

I interviewed recording & disk mastering engineer Clair Krepps (1918~2017) at his home 8/31/17. A Navy radio\radar technician, he became legendary at Capitol Records, MGM, Atlantic, and in 1965 his own Mayfair Studios. In the back corner above is an RIAA award for engineering Roberta Flack's *Killing Me Softly* and a Grammy for Nat King Cole's *A Christmas Song* by Mel Torné. On the floor is one of eight channel strips of the console he designed & built with brother Edgar for the industry-first 8-track recording studio. He was colleagues with Les Paul, Emory Cook, Normal Pickering, Walter Stanton, and other pioneers in the fledgling high-fidelity era of the 1950s. He served on Bergenfield NJ's Board of Education, and taught at the Jon Miller School of Recording Arts & Sciences [the author's brother]. One "trick" he taught was to mix to mono stereo sounds below 250Hz (others used 150Hz) for better reproduction of bass at home and on juke boxes without groove-hopping. "Not a *heavy compressor* guy," still his 1964 mastering of hit single "Do Wah Diddy" by Manfred Mann was widely regarded as the loudest 45 ever recorded! Criticized by colleagues for disrupting their "code of reasonable volume," it portended what has come to pass. Shortly after my interview, he turned 99, and received a life honor by the Audio Engineering Society (AES), which in 1948 he co-founded. In another month, this master of the microgroove was gone, but most deservedly not forgotten.

Acknowledgements, the author, and editor

Special thanks to editor R. A. "Bob" Bruner. Thanks to peer reviewers G. H. Aykroyd and H. S. Moscovitz. We four engineers have over two centuries of professional experience in analog & digital audio, video, and broadcasting. Many thanks to photo-microscopy mentor Mike Much, media advisor Kathy McAuley, and my spouse Nancy Desiderio for her love & tolerance. Also to Richard Steinfeld, author of *The Handbook for Stanton and Pickering Phonograph Cartridges and Styli*. *We stand on the shoulders of giants who came before us,* many more than referenced in [brackets] throughout. And thanks to our readers around the globe.

Robert E. (Robin) Miller III BSEE AES SMPTE BAS, author, is a pianist-arranger-bandleader, and *Peabody*-winning filmmaker. His 60+ years as an audio engineer began at age 14 completing the RCA Institutes radio correspondence course, then at 17 a FCC 1st Class (now "General") Radiotelephone License for his first job in Radio. He earned a BS in Electrical Engineering in 1967 (Lehigh University) where he won IEEE honors for his first scientific paper (a line array microphone). From 1970 his studio, Robin Miller, Filmaker Inc., with staff of seven (+20 part time) became the largest indie in Pennsylvania, making 300 film productions televised nationally, distributed globally, and recognized by 52 awards. In 1989 his was the first studio to electronically synchronize 16 & 35mm projectors, videotape, multi-track audio tape, and a digital server. And location digital audio recordings (for years recordist of the Greenwich Village Orchestra), big bands, pop\rock, and jingles. Since as Filmaker Technology, he is an adjunct professor, Patent-holder (full-sphere 3D reproduction), and a researcher who designs and publishes about audio and restores recordings and vintage equipment to play them.

R A Bruner, editor, has worked in broadcasting and professional sound since age 15. After being a literary magazine editor with a BA in linguistics, he earned a degree in engineering technology, and went on to design and built AM and FM radio facilities and TV production trucks. While at #3 market WTTW-TV11 and sister WFMT-FM98.7 in Chicago, he was responsible for conversion to a fully integrated digital HD production and broadcast facility renown for its national music & documentary programs for PBS and many others. Throughout his career, he's collected thousands of phonograph records, restored many vintage audio devices including turntables, and in retirement built a 1950s era radio museum.

Unless noted, images and text are property of the author, All Rights Reserved. No part of this book may be reproduced in any form, except for brief attributed quotation in reviews, without consent in writing from the author or his agent. Tradenames incidental herein are property of their owners.

Index

78rpm. See speed
acoustic music, 67, 73
align, 10, 13, 59, 118
Alignment. See align
all harmonics. See tone color, See distortion
anti-skating. See skating
archiving, 24, 28, 45, 66, 89, 119
artifacts. See distortion
azimuth error, 30
belt drive. See turntable
body. See cartridge
Breaking in, 50
calibration, 26
cantilever, 14, 15, 17, 20, 21, 24, 31, 33, 43, 45, 47, 48, 49, 50, 56, 60, 61, 62, 70, 77, 89, 119, 120, 132, 133, 139
capacitive load, 68, 74, 75, 130
cartridge, 8, 10, 11, 12, 14, 24, 38, 39, 45, 47, 48, 49, 58, 59, 60, 66, 68, 70, 71, 73, 74, 75, 76, 77, 79, 80, 81, 86, 88, 108, 111, 113, 114, 116, 118, 122, 123, 124, 126, 127, 128, 130, 131, 132, 133, 135, 136
C_{cable}. See C-load
CD, 8, 11, 24, 82, 83
chisel. See mastering
clean, 22, 65, 68, 136
cleaning, 11, 43, 65
C_{load}. See C-load
C-load, 6, 74
coloration, 68, 103, 133, 135, See distortion
compensation. See skating
compliance, 14, 16, 31, 33, 41, 47, 61, 63, 64, 80, 86, 89, 120, 131, 133, 136, 137
compression, 13, 99
conical. See spherical
contact, 12, 13, 16, 17, 18, 19, 20, 21, 22, 23, 24, 26, 28, 29, 31, 41, 42, 43, 44, 45, 48, 51, 53, 62, 81, 127, 130, 132, 133
correlated, 50, 54, 79

C_{preamp}. See C-load
cutoff. See f_3
direct drive. See turntable
direct-drive, 13
dirt, 22, 42, 43, 44, 65, 80
distortion, 6, 7, 8, 9, 10, 12, 13, 16, 17, 20, 21, 22, 23, 24, 26, 31, 33, 41, 43, 45, 50, 51, 53, 54, 57, 59, 62, 63, 67, 68, 70, 72, 73, 77, 78, 79, 80, 82, 86, 88, 91, 95, 99, 100, 101, 102, 103, 108, 115, 116, 117, 118, 119, 121, 130, 131, 132, 133, 134, 135
Dynagroove, 88
dynamics. See compression
elastomer. See cantilever
elliptical, 20, 21, 22, 23, 28, 29, 31, 33, 34, 35, 36, 37, 38, 39, 40, 41, 42, 43, 44, 45, 51, 53, 54, 62, 70, 81, 88, 90
erase, 23, 33
ET. See transcription
even-order, 51, 53, 54, 78, 99, 103, See tone color, See distortion
f_3, 8, 21, 22, 23, 24, 26, 28, 43, 99, 100, 135
fidelity. See distortion
friction. See skating
fully-automatic, 36
Gain controls, 30
gramophone. See phonograph
groove, 3, 7, 11, 12, 13, 14, 16, 17, 18, 20, 21, 22, 23, 24, 28, 29, 31, 33, 38, 41, 42, 43, 45, 49, 50, 51, 52, 53, 54, 57, 59, 61, 62, 63, 65, 66, 70, 77, 78, 79, 80, 81, 82, 84, 85, 86, 87, 88, 89, 99, 111, 115, 118, 119, 121, 131, 132, 133, 134, 135, 136, 139, 140
Groove speed, 22
habituation, 103
harmonic, 86, See distortion, See tone color
harmonic distortion (HD), 102
HD. See distortion, See harmonic distortion

headshell. See tonearm
heterodyne. See intermodulation
High fidelity, 102
high frequency, 17, 20, 21, 23, 24, 41, 43, 45, 54, 70, 72, 73, 75, 81
hill-and-dale. See vertical
hobbyist, 9
hyper-. See elliptical
idler (puck) drive. See turntable
IM. See intermodulation
imbalance, 73, 77, 79, 80
IMD. See distortion, See intermodulation distortion
impulse response. See resonance
inner. See groove
inner groove, 21, 23, 43, 54
interchangeable, 11, 33, 39, 73, 85, 120
intermodulation distortion, 56
kits. See maker project
Lateral, 49, 79
line-contact, 6, 11, 17, 20, 21, 22, 26, 33, 41, 43, 54, 67, 70, 88, 90
loading, 10, 70, 71, 74, 75, 86, 108, 111, 112, 116, 135, 139
loudspeakers, 77, 81, 91, 100, 101, 135, 139
LP. See microgroove
magnetic tape, 13, 58, 82, 83, 99, 120, 139
maker project, 65
mastering, 8, 17, 23, 26, 51, 73, 75, 80, 82, 83, 86, 87, 88, 115, 119, 120, 135, 139, 140
microgroove, 10, 20, 21, 33, 43, 44, 54, 82, 88, 120, 140
mis-tracking, 61, 63
modulation. See mastering
mono. See monophonic
mono switch, 53, 72, 115, 116
mono-ing, 73
monophonic, 21, 47, 50, 53, 54, 66, 70, 73, 80, 115, 116, 131, 134
moving coil (MC), 16, 47, 76
moving iron (MI), 16, 31, 44, 116, 135
moving magnet (MM), 16, 46, 116, 135
M-S, 52, 80
needle. See stylus, See stylus
Neumann Curve, 88
neutral. See distortion
new old stock (NOS), 20, 64
noise, 12, 23, 30, 57, 66, 72, 78, 80, 85, 86, 87, 90, 91, 93, 99, 100, 101, 102, 103, 108
normalizing. See compression
NOS (new old stock), 48
odd-order harmonics. See tone color, See distortion
offset angle, 59, 60, 62, 119, 123, 124, 132, 133, 134
overhang, 7, 59, 60, 119, 130, 131, 132, 133, 134, 139
perception, 12, 77, 87, 90, 92, 103
Pfanstiehl, 46, 47, 88, 139
phase, 29, 52, 53, 54, 56, 57, 73, 78, 79, 80, 86, 115, 116, 130, 133, 134, 135, 136
phono stage. See preamplifier
phonograph, 6, 9, 10, 12, 13, 14, 17, 22, 23, 29, 33, 39, 43, 47, 58, 65, 70, 73, 79, 90, 101, 116, 119, 135, 138, 139
pickup. See cartridge
pinch effect, 23, 53, 68, 79, 80, 99, See distortion
pink noise, 28, 30
Poid, 54, See all order harmonics
preamplifier, 10, 86
profile. See tip
protractor, 7, 19, 59, 60, 61, 89, 124
Quadrahedron. See line-contact
quadraphonic, 24
quality control, 17, 20, 26, 56, 75
racetracks. See contacts
record, 9, 10, 11, 12, 14, 17, 22, 23, 36, 49, 53, 59, 61, 63, 65, 70, 72, 76, 77, 79, 80, 82, 85, 86, 88, 99, 106, 113, 115, 117, 119, 121, 129, 130, 131, 133, 135, 139
Recording Industry Assn. of America standard (RIAA, 1954) curve. See RIAA
resonance, 10, 12, 61, 63, 64, 66, 68, 72, 73, 75, 89, 121, 123, 127, 131, 132, 135, 136
RIAA, 21, 23, 28, 64, 68, 70, 72, 80, 86, 87, 93, 114, 116, 135, 140
ripping. See archiving
rumble. See resonance
Samarium-Cobalt (SmCo). See moving magnet (MM)
scanning. See tracing

scratch, 22, 33, 43, 47, 65, 72, 133
semi-automatic, 36, 66
sensitivity. See cartridge
shape. See tip
shellac, 13, 23, 44, 49, 50, 63, 83, 88, 106, 119, 138
Shibata. See line-contact
sibilance, 12, 23, 26, 43, 45, 53, 79, 86, 91, 133
skating, 12, 19, 35, 61, 62, 63, 81, 99, 119, 120, 132, 133
skewed, 54, 56
SP, 43
spectrum analysis, 30
speed, 7, 20, 22, 23, 44, 45, 58, 62, 66, 68, 72, 84, 85, 89, 134, 135, 139
spherical, 17, 20, 21, 22, 23, 31, 33, 38, 39, 42, 43, 44, 45, 47, 48, 49, 51, 53, 54, 56, 62, 81, 86, 88, 120, 134
spinner. See turntable
standard play 78s & ETs. See wide groove
Stereohedron. See line-contact
stroboscope, 66
styli. See stylus
stylus, 3, 6, 7, 10, 11, 12, 13, 14, 15, 17, 18, 19, 20, 21, 22, 23, 24, 26, 28, 29, 30, 31, 32, 33, 38, 39, 40, 41, 42, 43, 45, 46, 47, 48, 49, 50, 51, 52, 54, 58, 59, 60, 61, 62, 63, 64, 65, 66, 68, 70, 71, 73, 77, 78, 79, 81, 87, 117, 119, 120, 122, 123, 127,129, 130, 131, 132, 133, 134, 136, 139
Stylus Rake Angle. See align
Styrene, 63
sum & difference. See intermodulation
sum & difference tones. See intermodulation distortion
THD. See distortion, See distortion, See distortion
timbre. See tone color, See tone color, See tone color
tip, 6, 8, 13, 14, 16, 17, 18, 19, 20, 21, 22, 23, 24, 31, 37, 43, 44, 45, 48, 49, 50, 51, 54, 56, 59, 60, 61, 62, 65, 82, 84, 119, 120, 124, 126, 130, 134

tone color, 7, 8, 13, 29, 74, 78, 86, 89, 90, 91, 93, 94, 101, 102, 107, 108, 115, 116, 135
tonearm, 6, 8, 10, 12, 14, 21, 22, 58, 59, 60, 61, 62, 63, 64, 65, 66, 68, 70, 71, 72, 89, 111, 112, 114, 115, 116, 117, 119, 121, 122, 128, 129, 130, 131, 132, 133, 134, 135, 136, 139
total harmonic distortion, 61
tracing, 12, 20, 21, 22, 23, 24, 26, 28, 29, 33, 41, 45, 51, 62, 70, 81, 134
tracking. See align
tracking force, 12, 22, 23, 28, 32, 61, 81, 120, 125, 129, 132
transcription, 6, 10, 60, 61, 62, 64, 66, 82, 117, 119, 120, 121, 131
transparency. See distortion
turntable, 7, 9, 10, 12, 13, 14, 21, 28, 36, 38, 40, 58, 63, 65, 66, 71, 76, 77, 82, 84, 85, 87, 111, 112, 114, 115, 117, 118, 119, 120, 121, 122, 128, 129, 131, 133, 135, 137, 141, See phonograph
ultrasonic, 24, 56, 65, 75, 83, 88
uncompressed. See compression
uncorrelated, 79
Universal P-mount, 39
variable reluctance. See moving iron (MI)
vectorscope, 51
verisimilitude, 95, 100, 103, 120, 138
vertical, 20, 21, 22, 23, 26, 32, 38, 41, 45, 50, 52, 53, 54, 77, 78, 79, 80, 82, 84, 87, 88, 89, 108, 113, 116, 120, 125, 129, 131, 136
Vertical Tracking Angle. See align
Victrola, 49, 82, 106
vinyl. See record
warp, 51, See resonance
wear, 8, 10, 12, 13, 17, 18, 20, 21, 22, 23, 24, 28, 29, 33, 38, 41, 42, 43, 45, 47, 48, 49, 51, 62, 65, 70, 124, 132, 134
wide-groove. See SP
wow & flutter, 66

A long time ago in a galaxy far away...

...earth was invaded by an alien medium – Radio. Invisibly it entered their homes, then their cars. Before, the earthlings' home entertainment had been the Victrola, the player piano or books that required reading. Their news was delivered on paper a day late, or on reels at the cinema a week after it happened. But this Radio spoke the news "live," almost as it happened. It told jokes and stories, with laughter and sound effects, and sang songs in many languages. Families gathered around its altars of wood with glowing dials, warming themselves over its light-saber tubes. Little did they know that for every disembodied android they heard performing, there were ten others writing the words and songs, producing the programs, selling sponsors, and maintaining the starships. Some became famous planet-wide, or at least around town. And the Age was called Golden. Then a revolt: robots began to program the airwaves; many creative and technical earthlings were put out of work. And only 40 songs were allowed to be played. But a new alien form saved the day – with pictures. Now earthlings had 500 channels, but some could still find nothing to watch. And few paid any attention to the radio, now that it was just noise. Until still another alien medium emerged – the electronic book – that told stories on screens as small as your hand. One with the title "American Radio Then & Now" kept the history of Radio alive well beyond the Golden Age. And earthlings Googled its title, and downloaded it from a great river of data called Amazon, and gave the eBook stars. And those who were old remembered, and those who were young marveled at what once was. Bringing them together once again.

Also by Robin Miller – The stories, the pioneering technology, the unseen practical jokes among the hundreds of thousands who worked on-air & behind the scenes in local Radio in The Golden Age. Free sample at www.amazon.com/dp/**B0141JRPN0**

Acknowledgements, the author, and editor

www.ingramcontent.com/pod-product-compliance
Lightning Source LLC
Chambersburg PA
CBHW041430300426
44114CB00002B/22